Problems in Engineering Graphics and Design
SECOND EDITION

André M. Skaff
University of Ottawa
Ottawa, Ontario, Canada

www.kendallhunt.com
Send all inquiries to:
4050 Westmark Drive
Dubuque, IA 52004-1840

Copyright © 1983, 2010 by Kendall Hunt Publishing Company

ISBN 978-0-7575-7236-4

All rights reserved. No part of this publication may be reproduced, stored in a retrieval system, or transmitted, in any form or by any means, electronic, mechanical, photocopying, recording, or otherwise, without the prior written permission of the copyright owner.

Printed in the United States of America
10 9 8 7 6 5 4 3

Table of Contents

INTRODUCTION 1

ACKNOWLEDGEMENT 2

CHAPTER 1 - LETTERING 3

    Exercises 4

CHAPTER 2 - USE AND CARE OF INSTRUMENTS AND GEOMETRICAL CONSTRUCTIONS 6

    I. Use and Care of Instruments 6

    II. Geometrical Construction 6

    Exercises 8

CHAPTER 3 - SKETCHING 13

    I. Sketching a Straight Line 13

    II. Sketching a Circle 13

    III. Sketching an Ellipse 14

    Exercises 16

CHAPTER 4 - ORTHOGRAPHIC PROJECTION 22

    Exercises 25

CHAPTER 5 - AUXILIARY VIEWS 34

    Exercises 35

CHAPTER 6 - SECTIONS 38

    Types of Sections 38

    Conventions 38

    Exercises 40

CHAPTER 7 - DIMENSIONING 46
I. Rules of Technique 46
II. Rules of Locating Dimensions... 46
Exercises 48

CHAPTER 8 - PICTORIAL DRAWINGS 52
I. Axonometric Projection... 52
II. Oblique Projection 53
III. Perspective 54
Exercises 55

CHAPTER 9 - FASTENERS 62
Standard Bolts 63
Screws 63
Rivets 63
Exercises 64

CHAPTER 10 - DIMENSION LIMITS, TOLERANCE AND ALLOWANCE 65
Exercises 67

CHAPTER 11 - DESCRIPTIVE GEOMETRY 68
Exercises 70

CHAPTER 12 - INTERSECTIONS AND DEVELOPMENT OF SURFACES 78
I. Intersections 78
II. Development of Surfaces 78
Exercises 79

CHAPTER 13 - INTRODUCTION TO AUTOCAD 85
- AutoCAD interface 85
- Screen Background Colour 86
- Drawing Units 86
- Drawing Limits 86
- Zoom and Pan 86
- Draw Toolbar 87
- Modify Toolbar 90
- Layers 92
- Dimension Toolbar 93
- Page Layout 95
- Exercises 96

INTRODUCTION

This book is designed to accompany a basic course in Engineering Graphics. The lessons and problems are prepared and arranged in such a way to benefit students who want to acquire a good understanding of Engineering Graphics methodology. The sequence of the different chapters is arranged in such an order that the author believes facilitates the acquisition of the material by the reader.

As is the case in all aspects of modern technology, the advent of the high-speed computers has greatly influenced drafting technology and nowadays, drawings of complex problems are made easier and faster by the development of Computer-Aided Design & Drafting (CADD) systems. Much software was developed in this respect, but one that is widely and more commonly used than the others is AutoCAD. However, all of these software have similar principles and objectives which are to make the drawings of more complex work more precise and made in much less time than manual ones will take. Better still, drawings will be modified in timeless manner and design information is conveyed efficiently to a design team that may be in different locations. This new development represents a major breakthrough in the design of projects which vary from simple to more complex ones. For the above reasons, a section of this book is dedicated to AutoCAD and how it is used. The development of computers and software, although extremely important, do not remove the fundamental need for Engineers and Drafters to acquire good and basic knowledge of drafting principles. After all, they are the ones who will be feeding the information to the computers and themselves are the ones who will be interpreting the drawings and explaining them.

Engineering Drawings should be looked upon, with or without computers, as the language by which engineers communicate their ideas visually. This graphical language may be defined as the study by which three-dimensional objects are represented on two- dimensional surfaces as the plane of a drawing paper or screen of a computer. This is an international language in the sense that a drawing made in one country can be interpreted and implemented in another.

This book is arranged in such a manner that at the beginning of every chapter or subject, a brief introduction to the topic is given and a reference to the techniques that may be used in solving the problems relating to this chapter or topic. Because of its adoption by many countries, the SI system of measurements is used in most of the problems contained in this book.

I hope that this book, in its revised edition, will meet the expectations and needs of all those who are involved in teaching and learning Engineering Graphics.

The Author

ACKNOWLEDGEMENT

The present edition of this book was made possible with the help of Mr. Mansour Navidpour, a PhD candidate in the department of Civil Engineering at the University of Ottawa. His good knowledge and computer skills in the domain of Graphics helped in the production of this copy and he deserves my full gratitude and appreciation.

CHAPTER 1

LETTERING

- Lettering rather than script is used on engineering drawings.
- There are three styles of letters:
    - Gothic
    - Roman
    - Old English

    Gothic is the one most commonly used by engineers and architects so it is the one which is illustrated in this book.
- The Gothic letter could be:
    - Single stroke, double stroke or filled in
    - Vertical or slant
    - Capital or small
- The size of lettering is the height of the Capital letter in a word. For the sake of legibility and clearness, the size should never be less than 3 mm or 1/8 inch. The height of a small letter should be 2/3 that of the capital.
- Numerals have the same height as that of capital letters while fractions are 1 ½ times that height.
- Letters are always written with the help of guidelines.
- A rather dark pencil should be used in lettering. F pencil will be the best but H and 2H are used as well.
- All elements of letters should be done with one stroke of the pencil. All strokes should be done from top to bottom and from left to right.
- To obtain good lettering technique, one should maintain uniformity of shape, style, size, slope, weight and spacing.

NAME _____
SECTION . _____

A B C D E F G H I J

K L M N O P Q R S T

U V W X Y Z

1 2 3 4 5 6 7 8 9 0

a b c d e f g h i j k l m

n o p q r s t u v w x y z

1 2 3 4 5 6 7 8 9 0 3$\frac{1}{4}$ 7$\frac{1}{2}$

AN ENGINEER MAY BE DEFINED AS THE PERSON WHO DESIGNS AND BUILDS DEVICES OR FACILITIES THAT ARE BENIFICIAL TO MANKIND.

Copy, freehand, the above letters and numbers in between the guidelines below each line to form other rows of uniformly spaced letters and numbers. Repeat the last sentence once.

NAME
SECTION.

A B C D E F G H I J

K L M N O P Q R S T

U V W X Y Z

1 2 3 4 5 6 7 8 9 0

a b c d e f g h i j k l m

n o p q r s t u v w x y z

1 2 3 4 5 6 7 8 9 0 3$\frac{1}{4}$ 7$\frac{1}{2}$

AN ENGINEER MAY BE DEFINED AS THE PERSON WHO DESIGNS AND BUILDS DEVICES OR FACILITIES THAT ARE BENIFICIAL TO MANKIND.

Copy, freehand, the above letters and numbers in between the guidelines below each line to form other rows of uniformly spaced letters and numbers. Repeat the last sentence once.

CHAPTER 2

USE AND CARE OF INSTRUMENTS AND GEOMETRICAL CONSTRUCTIONS

I. Use and Care of Instruments

An engineering student should know how to use the instruments he or she is working with properly in order to save time and secure good and accurate results. Also, he or she should know how to take care of these instruments and keep them in good shape.

Some of the precautions and recommendations which will help the user keep the instruments in good shape and thus secure good and proper drawings are the following:

- Never use the scale as a sharp edge to draw straight lines. The triangles and the T-square are used for that purpose.
- The T-square and triangles should not be used as guides to cut papers by a razor blade or a knife.
- Do not use the T-square as a walking stick or as a hammer.
- Fix the drawing paper in a convenient place in the lower left hand corner of the drafting table.
- Keep your triangles clean by washing them frequently.
- Do not open the bow compass by spreading away its two legs by force.
- Use the brush to remove the erasings off the paper; do not use the palm of your hand.
- It is preferable to use the divider in copying distances rather than the scale.
- Turn the pencil slightly in your fingers while drawing straight lines. This will secure uniform thickness of the lines.
- Use different hardnesses of leads for drawing different lines:
  - For visible lines, use F or H pencils
  - For hidden (invisible) lines, use 2H pencils
  - For center lines and dimension lines, use 4H pencils
  - For construction lines and guidelines, use 6H pencils.

II. Geometrical Constructions

Using drawings instruments, an engineer or draftsperson must know how to execute the following geometrical constructions:

- Bisect a line.
- Construct a line parallel to another through a given point.
- Construct a line perpendicular to another through a given point.
- Construct a line perpendicular to another at one of its extremities.
- Divide a line into a given number of equal parts.

- Layout an angle.
- Bisect an angle.
- Construction of polygons, circles, ellipses, etc.

1. Construct the circle circumscribing triangle ABC by using a compass and a straight edge only.

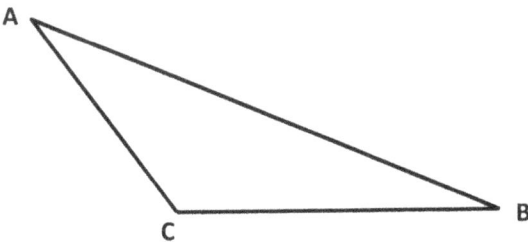

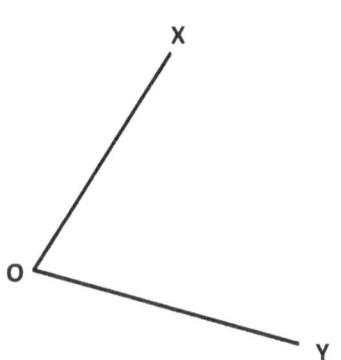

2. Construct line PQ, 50 mm long, making with QR an angle equal to half of the given angle XOY. Use a compass and a straight edge only.

TITLE:	
DRAWN BY:	
CHECKED BY:	
SCALE:	DATE:
SECTION:	DRAWING No.:

1. Construct, on one side of MN, line OM, 30 mm long and perpendicular to MN at M by using the 3, 4, 5 method.

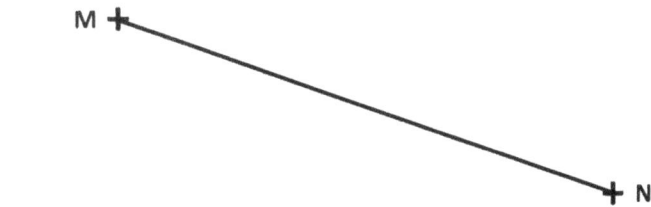

2. Construct a regular hexagon with side AB by using the triangles only.

TITLE:	
DRAWN BY:	
CHECKED BY:	
SCALE:	DATE:
SECTION:	DRAWING No.:

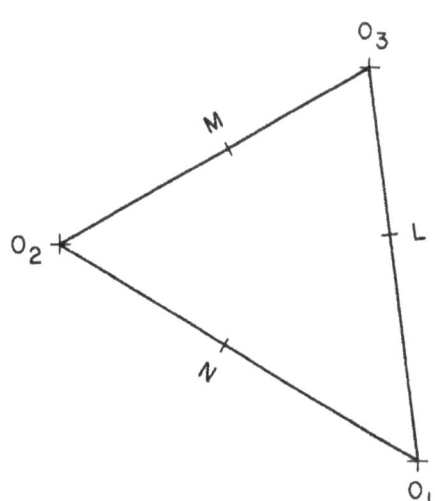

O_1, O_2 and O_3 are the centers of three circles tangent to each other at points M, N and L. Draw these circles and their common exterior tangents.

TITLE:	
DRAWN BY:	
CHECKED BY:	
SCALE:	DATE:
SECTION:	DRAWING No.:

GEOMETRICAL CONSTRUCTION

Plot the parabola $y = 2x^2 + 3$ for values of x varying from -5 to +5 units, on a graph paper from the back of the book. Use one centimetre on the x-axis to represent one unit and one centimetre on the y-axis to represent 5 units. Use the FRENCH CURVE to join the points of this parabola.

HYDROLOGY PROBLEM

The average monthly precipitation (rainfall) and runoff for a certain area is the following:

Month	Precipitation (P)	Runoff (Q)
	mm	mm
January	76.4	63.9
February	87.1	43.3
March	28.5	46.2
April	43.6	30.9
May	84.3	20.1
June	137.8	24.5
July	81.6	5.4
August	175.7	26.3
September	9.2	2.8
October	46.4	1.7
November	47.4	3.3
December	35.8	2.1
	857.8	270.5

Plot these values on a plane system of coordinates. The horizontal axis will represent the date and the vertical one, the amount of precipitation or runoff. Join these points in the right order, FREEHAND or with the FRENCH CURVE, to form two curves, one for the precipitation and one for the runoff.

CHAPTER 3

SKETCHING

To sketch is to draw freehand and an engineer should know how to do this properly. Sometimes an engineer, whether he/she is in the field or in the office, is called upon to explain something about a certain project. This can be done by means of a freehand diagram (called a sketch) which requires much less time than drawing with instruments. Therefore, an engineer should be capable of making clear and understandable sketches; otherwise it will confuse and mislead the people he/she is explaining to.

Papers used in sketching could be of either the coordinate type, where dimensions could be easily estimated, or of the blank type. Soft pencils such as F, H, or 2H are used in this respect.

Some of the techniques followed in sketching are the following:

I. Sketching a Straight Line
a) Mark down the end points of the straight line.
b) Hold the pencil in a normal writing position and place it at the left point in a horizontal line and at the far point in a vertical line.
c) By keeping your eyes fixed at the other end, trace a very light line by moving the pencil towards that end.
d) Go over that line and mark it down with one stroke to make it to the required weight.

If the beginner experiences some difficulty in sketching a certain line (horizontal, vertical or inclined) while he/she is good in another, he/she could turn the paper around until that line is in the position that suits his/her convenience.

II. Sketching a Circle
A circle could be swung by a fast movement of the hand. This is a very rough method and it takes a lot of practice to obtain good results. However, the following methods are found very useful:

a) The Trammel Method:
- Mark down two points, say O and C, on the edge of a piece of paper such that the distance between them is equal to the radius of the required circle.
- Place one of the points, say O, at the center of the circle.
- Keep point O fixed at the center and by rotating the paper around, mark down points at C.

- Join these points freehand to obtain the required circle. The more points one takes, the more accurate results one gets.

b) The Circumscribed Square Method:
- Sketch out a light square of side equal to the diameter of the circle and mark down the mid-points of the sides.
- Sketch out the arcs tangent to the sides at their mid-points in proper direction and curvature.
- Join these arcs to obtain a smooth circle inscribed within the square and go over it to make it darker and to improve its weight.

c) The Semi-Mechanical Method:
- Hold two pencils between the thumb, the fore finger and the middle finger of your hand so that they form a shape that looks like a compass. Open the pencils to the required radius.
- Fix the inner pencil on the center of the required circle and let the other one just touch the paper.
- Rotate the paper around the inner pencil which acts as a pivot and the other pencil traces the circle.

After practicing a while, the last method could give you good and fast results.

III. Sketching an Ellipse

As for a circle, an ellipse could be swung by a fast movement of the hand. This again is a very rough method and it takes a lot of practice to obtain good results. However, the following methods are found very useful:

a) The Trammel Method:
- Mark down three points, say A, B and C, on the edge of a piece of paper such that distance AB is equal to half of the major axis of the ellipse and BC is equal to half of the minor axis of the ellipse. B is between A and C.
- Place the end points A and C on the two axes of the ellipse.
- Rotate the paper such that A and C will remain each on any axis and mark down points at point B.
- Join these points freehand to obtain the required ellipse.

b) The Circumscribed Rectangle Method:

- Sketch out a light rectangle with sides equal respectively to the major and minor axes of the required ellipse.
- Sketch out arcs with proper curvatures tangent to the sides of the rectangle at their mid-points.
- Connect these arcs together to obtain a smooth curve and go over it to make it darker and to improve its weight.

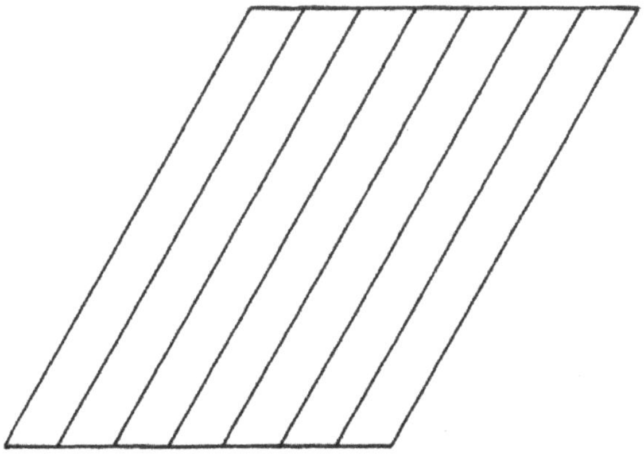

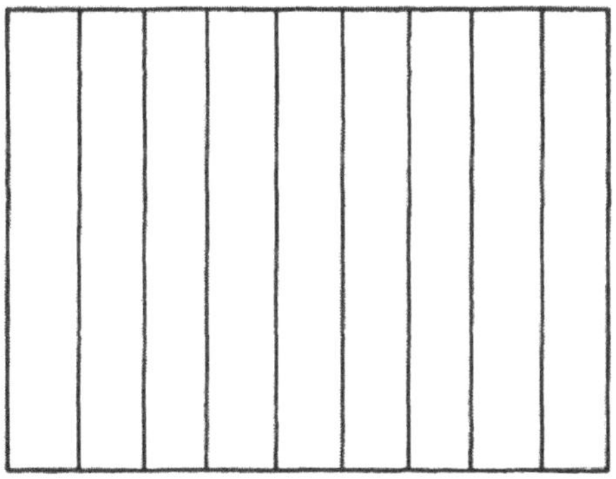

Sketch, to the same size, the above figures in the space to the right.

TITLE:	
DRAWN BY:	
CHECKED BY:	
SCALE:	DATE:
SECTION:	DRAWING No.:

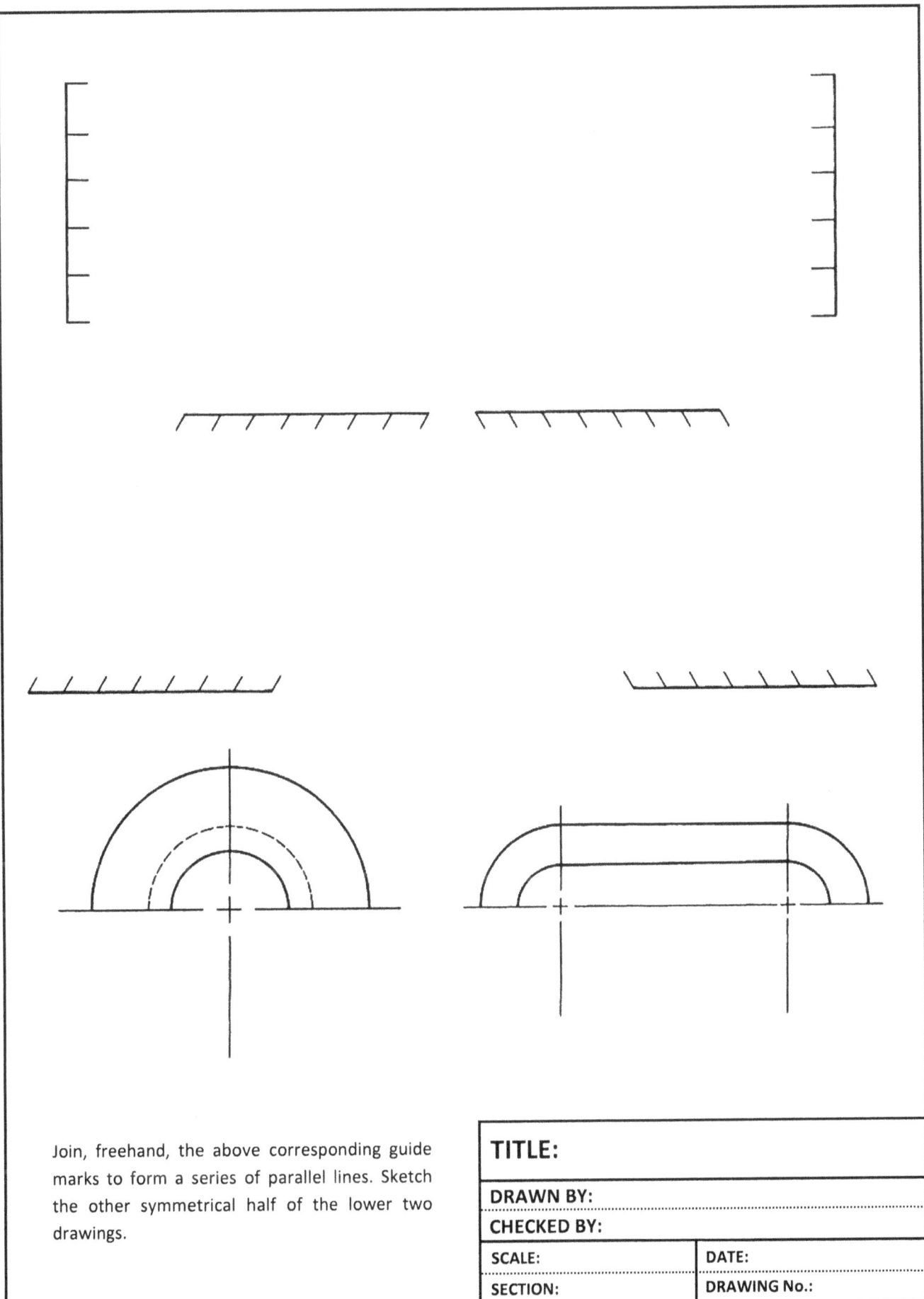

Join, freehand, the above corresponding guide marks to form a series of parallel lines. Sketch the other symmetrical half of the lower two drawings.

TITLE:	
DRAWN BY:	
CHECKED BY:	
SCALE:	DATE:
SECTION:	DRAWING No.:

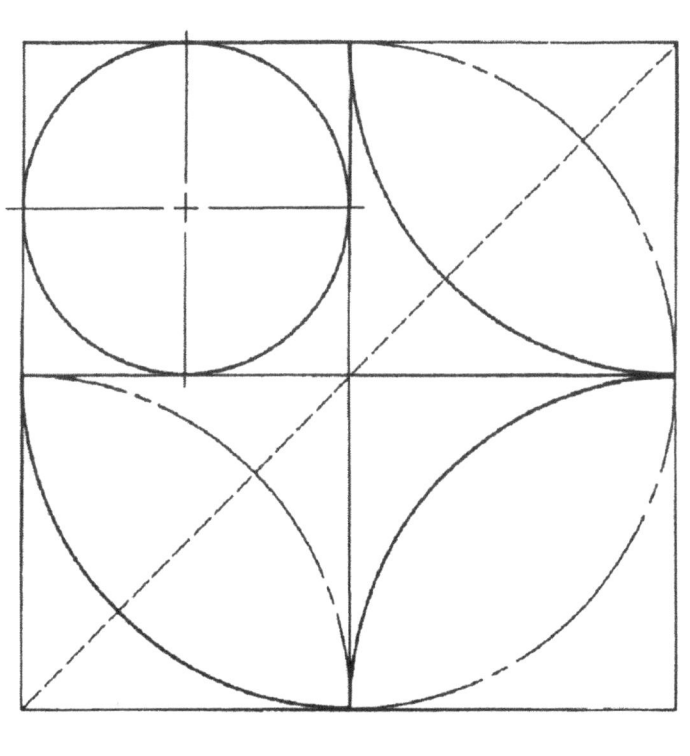

Sketch the above drawing in the space below it.

TITLE:	
DRAWN BY:	
CHECKED BY:	
SCALE:	DATE:
SECTION:	DRAWING No.:

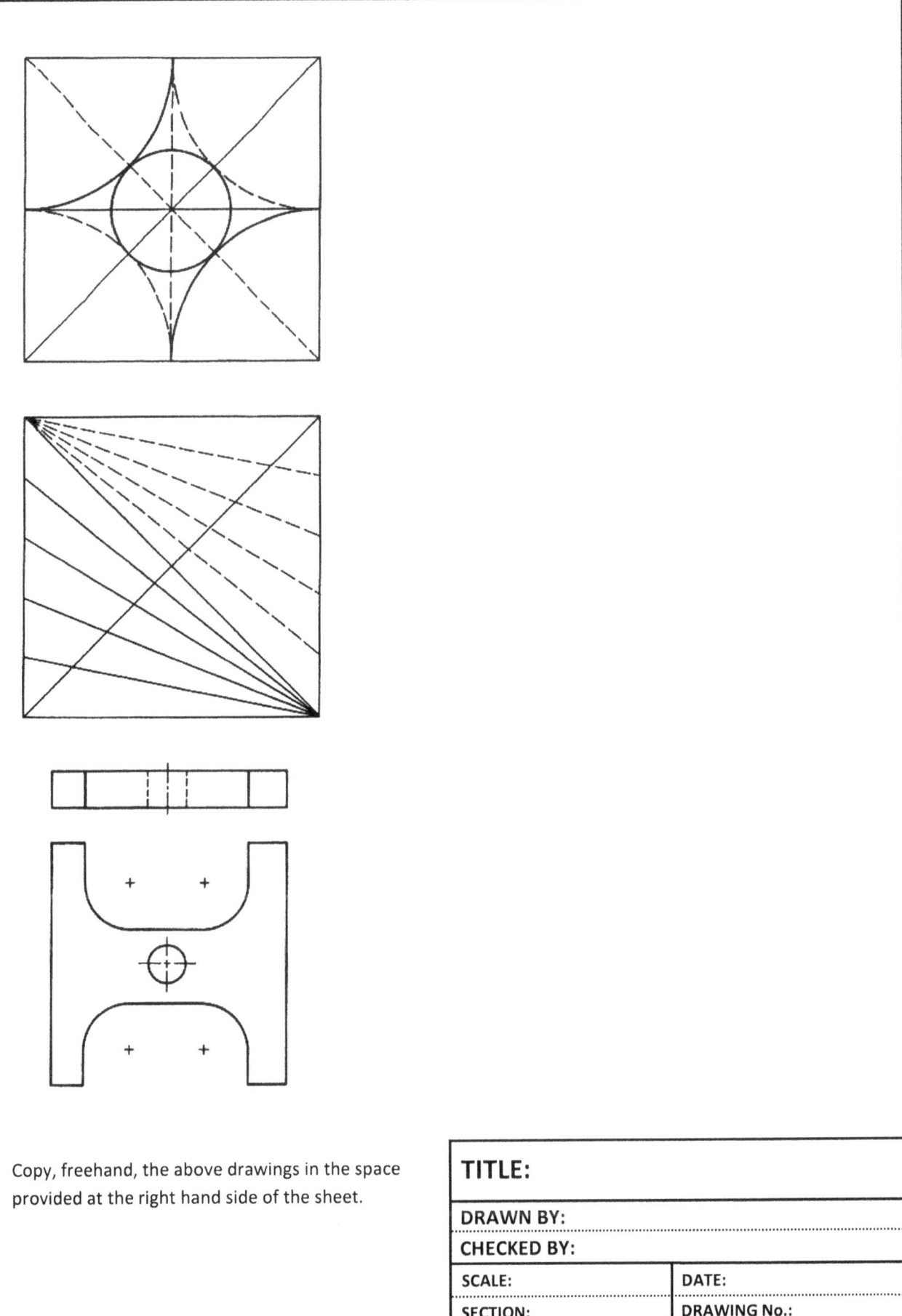

Copy, freehand, the above drawings in the space provided at the right hand side of the sheet.

TITLE:	
DRAWN BY:	
CHECKED BY:	
SCALE:	DATE:
SECTION:	DRAWING No.:

The topographic data obtained in the field for the drawing of a contour map of a certain area is shown below. Join, freehand, all points of the same kind and in the right order to form a series of concentric loops similar to the one shown by the example. These loops are called contour lines. (A contour is a line that connects points of equal elevation).

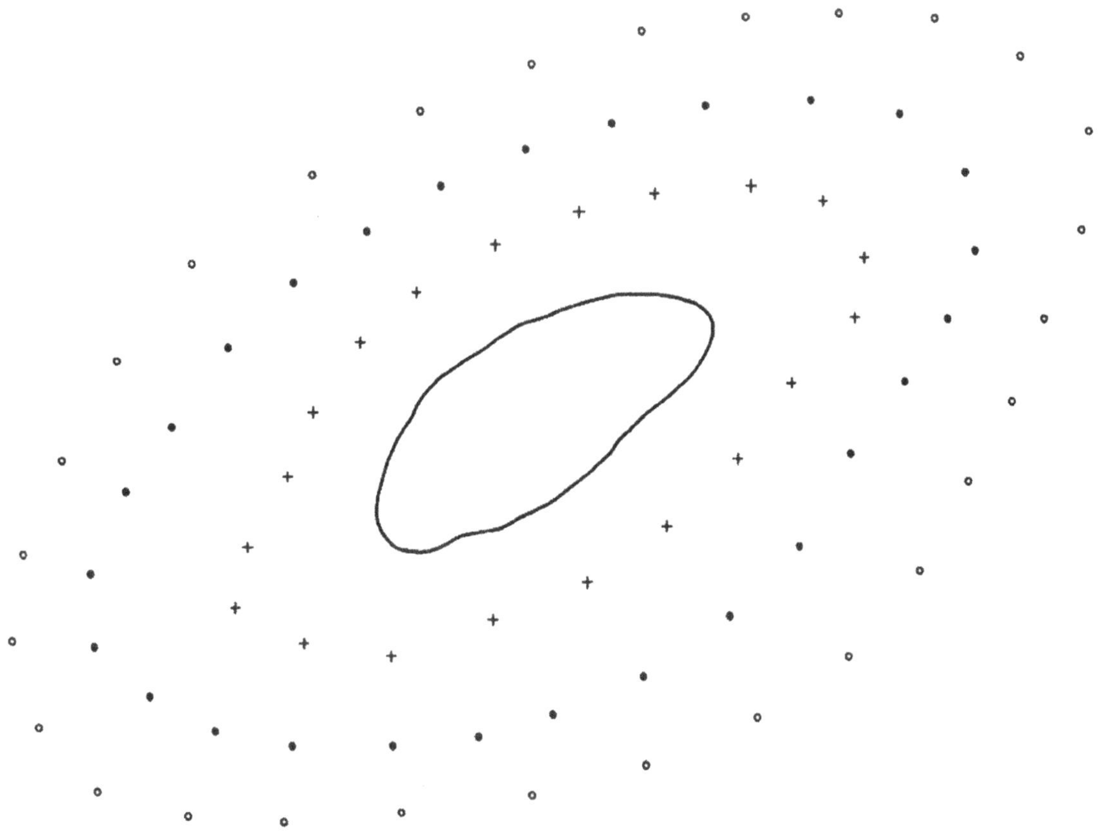

TITLE:	
DRAWN BY:	
CHECKED BY:	
SCALE:	DATE:
SECTION:	DRAWING No.:

SKETCHING PROBLEM

A rectangular field, 200 m long and 100 m wide, has two semicircles, one at each narrow end tangent to the longitudinal sides of the field. A running track, 20 m wide runs parallel to the sides on their periphery.

Draw freehand the field and the running track on a blank sheet of paper from the back of the book with the long dimension parallel to the long dimension of the sheet.

CHAPTER 4

ORTHOGRAPHIC PROJECTION

Orthographic projection is a method by which a three-dimensional object is represented on a two-dimensional surface. This is done by projecting all points of a certain object on various planes of projection by means of parallel lines which are perpendicular to these planes. It is the image of an object outlined on a plane surface by looking through that surface from an infinite point of sight. Representation of objects by the method of orthographic projection is the most commonly used one in engineering.

In orthographic projection, three principal planes mutually perpendicular to each other are used. These are: Horizontal (H), Vertical (V) and Profile (P) planes of projection. The horizontal and vertical planes divide the space around them into four quadrants and the object could be placed in any of them. In the United States, Canada and some other countries of the world, drawings are made with the object placed in the third quadrant; that is, underneath the horizontal plane and behind the vertical one. This kind of projection is referred to as "Third Angle Projection". The profile plane, which is also a vertical plane from the side, could be placed either to the right of the object or to its left.

The object could then be considered as placed within a transparent box with the main sides parallel to the three principal planes of projection. By looking from an infinite distance through the various faces of the box, different images of the object are seen. These images are referred to as "views" and thus six principal views are outlined on the six faces of the box. Now, if the faces of the box (with the views on them), were unfolded and rotated in such a way that they all lie in the same plane as the front face, then the 3-dimensional object would have been transferred into 2-dimensional views that fully describe the object. Rotation is being done away from the object. By studying these views, it is found that each opposite pair shows the same features and dimensions of the object but in the reverse order. Therefore, three out of the six views could be eliminated because they do not reveal any additional information about the object. Use only the top, front and right side views unless one of the others is needed. In many cases the left side view is used in lieu of the right side view. It is worth noting that each dimension of the object appears twice in these three views. For instance, the length of the object appears in the front and top views, the width or depth appears in the top and side views, and the height or elevation appears in the front and side views.

Some objects require three views to be fully represented or described (three-view drawings), some require two views (two-view drawings), and some, a single view would be enough if the third dimension is given by a note (one-view drawings).

In the views, lines that represent visible features of the object are drawn as solid continuous lines (**visible lines**), lines that represent hidden features are drawn as dashed lines (**invisible lines**), and **centerlines** are drawn to all circular and symmetrical parts. It is worth mentioning that visible lines have priority over invisible lines and centerlines if they happen to coincide in a certain view. In other words, the visible line is drawn and the others are eliminated in the parts

where they coincide. Similarly, invisible lines have priority over centerlines if they happen to coincide.

In selecting the views in multi-view drawings, the following rules may be followed:

1. The object should be placed or imagined in its natural position before the views are drawn.
2. Select the front view to be the one which is most descriptive of the object.
3. After selecting the front view, the top and side views will be automatically selected.
4. The proper side view to be drawn should be that one which shows fewer invisible lines.

It is a good practice to place the front and top views in vertical alignment with each other and the front and side views in horizontal alignment with each other. This arrangement will help correlate the views and transfer dimensions from one view to another. In making a multi-view drawing, it is advisable to follow the following steps:

1. Choose the number and arrangement of views.
2. Make a freehand layout of the areas for each view. From this, determine the spacing of the views.
3. Make an accurate mechanical layout of the outlines of the views. (Beginners are advised to use reference lines). Then, add the main centerlines and the centerlines of the details.
4. Draw the circular parts.
5. Draw the straight lines.
6. Locate intersections and irregular curves.
7. Clean up the drawing; that is, take out excess lines, construction lines and reference lines.
8. To improve the technique, go over all the lines to make them heavier.
9. Put in title and scale.

Interpretation of A Drawing

So far, we have learned how to draw the views if the object is given. If the views are given, can we read them or interpret them to establish what the object looks like? That is, can we visualize the object from the views?

Maybe this is the most difficult part of this lesson because there are no fixed rules that could be followed. As a matter of fact, a good solution to this problem is to use our imagination in putting the various parts of the object together to obtain a correct pictorial view. Maybe one can start by sketching a rectangular box having dimensions equal to the maximum dimensions of the views. Then, sketch the views in pictorial form in the respective faces of the box. Afterwards, start taking out the parts that are not needed. Certainly, before starting to analyze

a certain drawing, the reader should understand what the different lines (straight or curved) which make up the views might represent. For example, a straight line may represent one of the following things: intersection of two planes, the edge view of a plane surface or the outside line of a curved or a cylindrical object.

It is a good practice, after visualizing the object and maybe sketching its pictorial view, to go back to the views and check if they conform to each other.

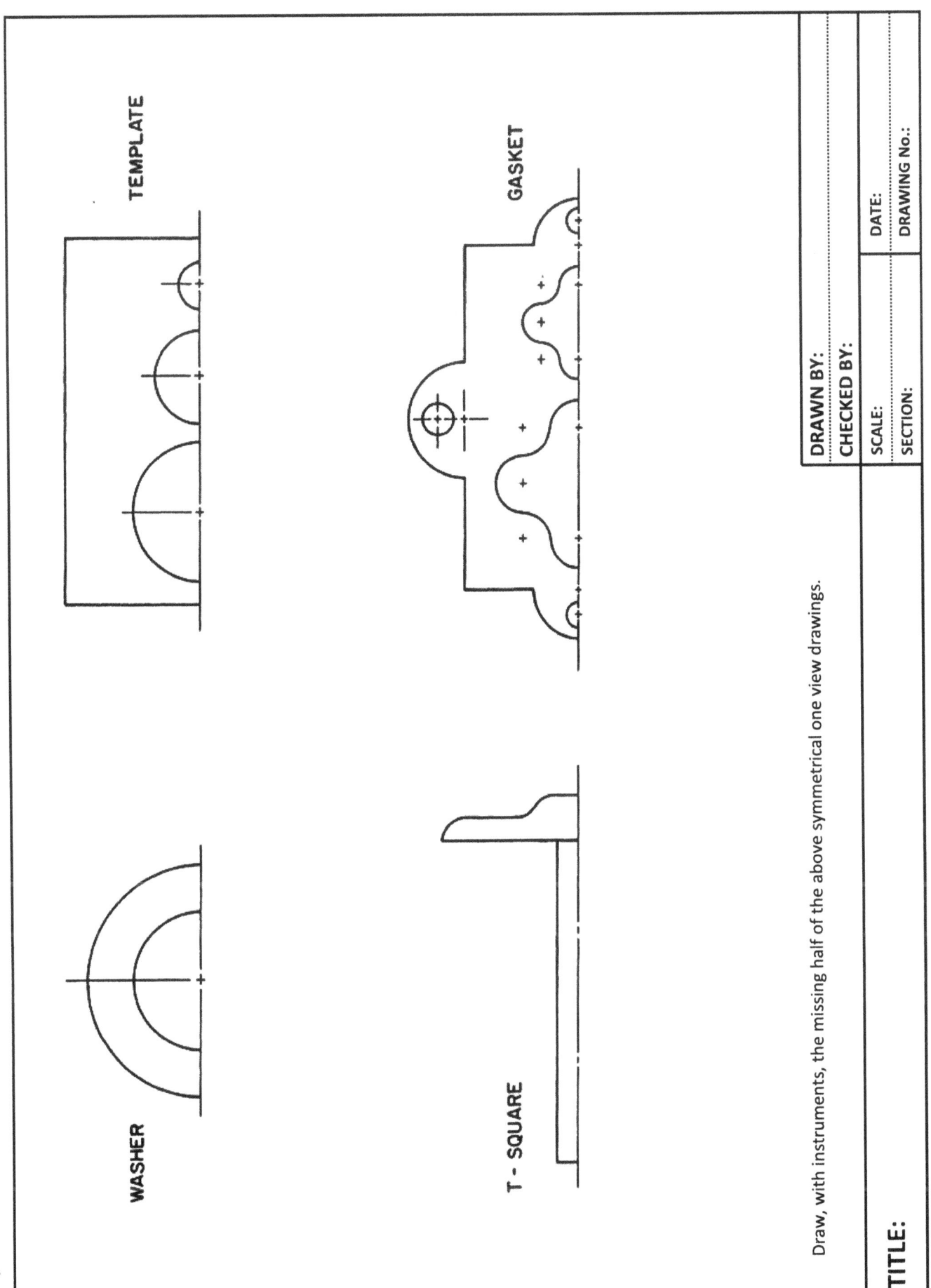

Draw, with instruments, the missing symmetrical half of the above two drawings.
Locate all centers of all circles or arcs.

TITLE:

DRAWN BY:
CHECKED BY:

SCALE:
SECTION:

DATE:
DRAWING No.:

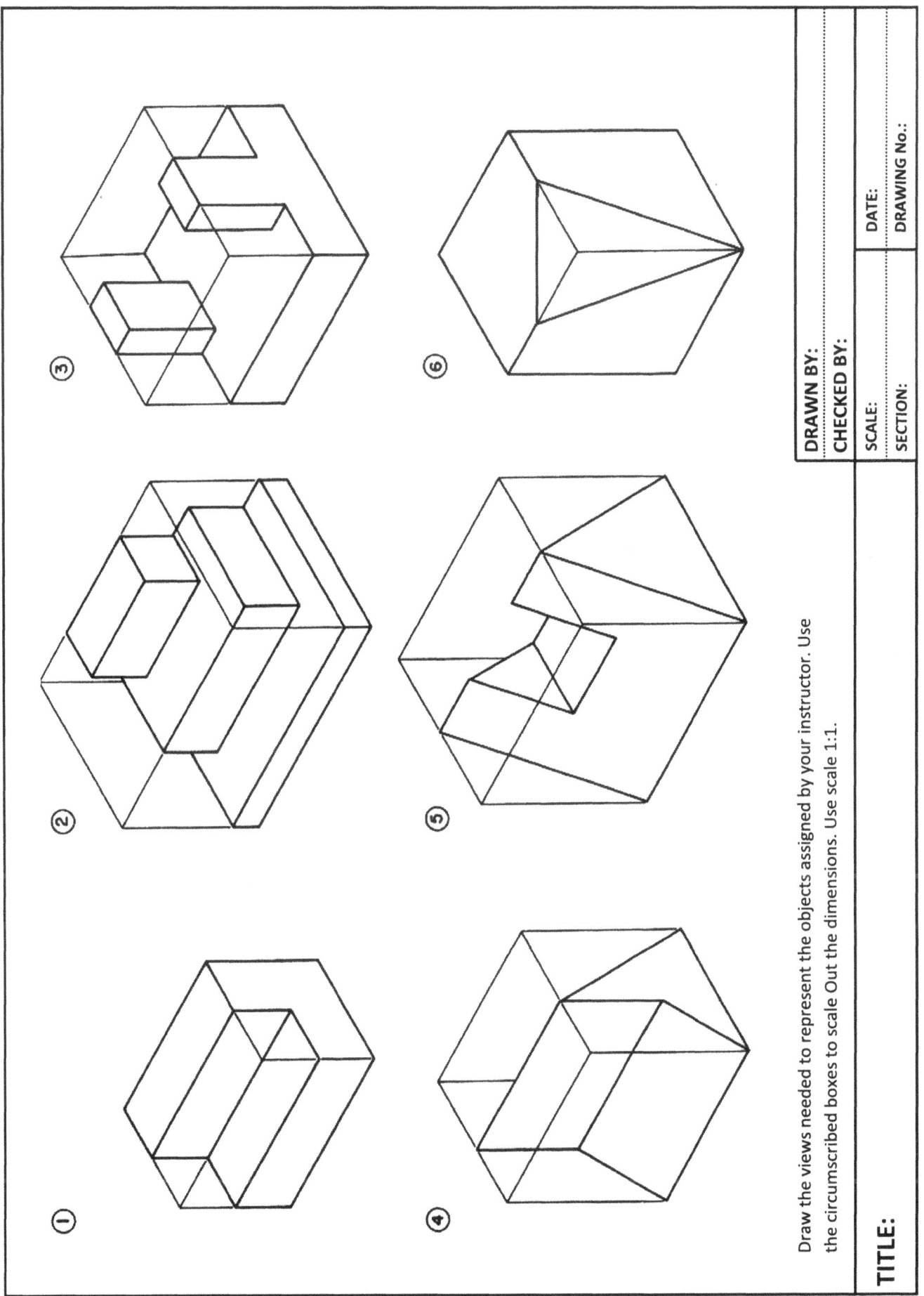

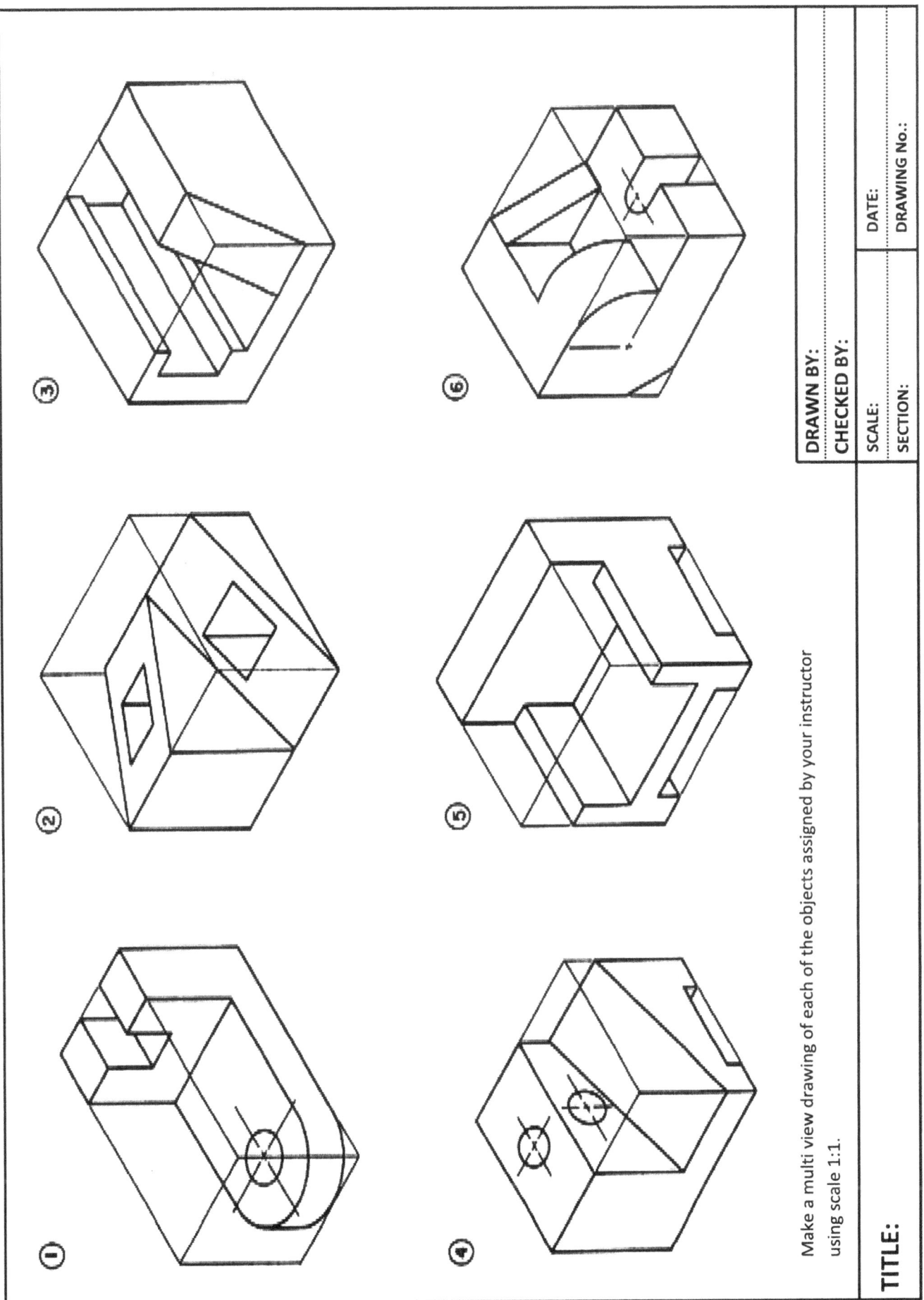

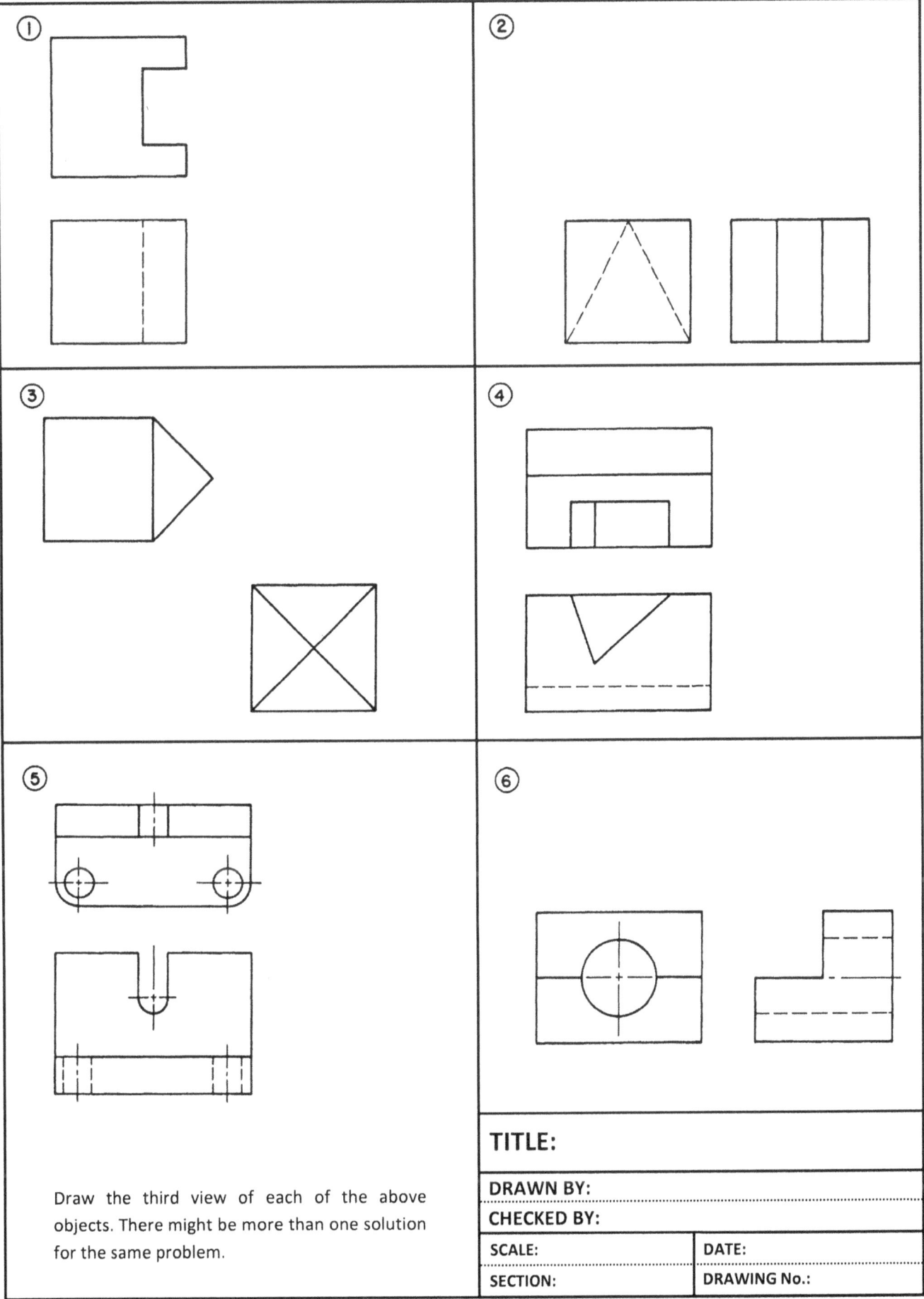

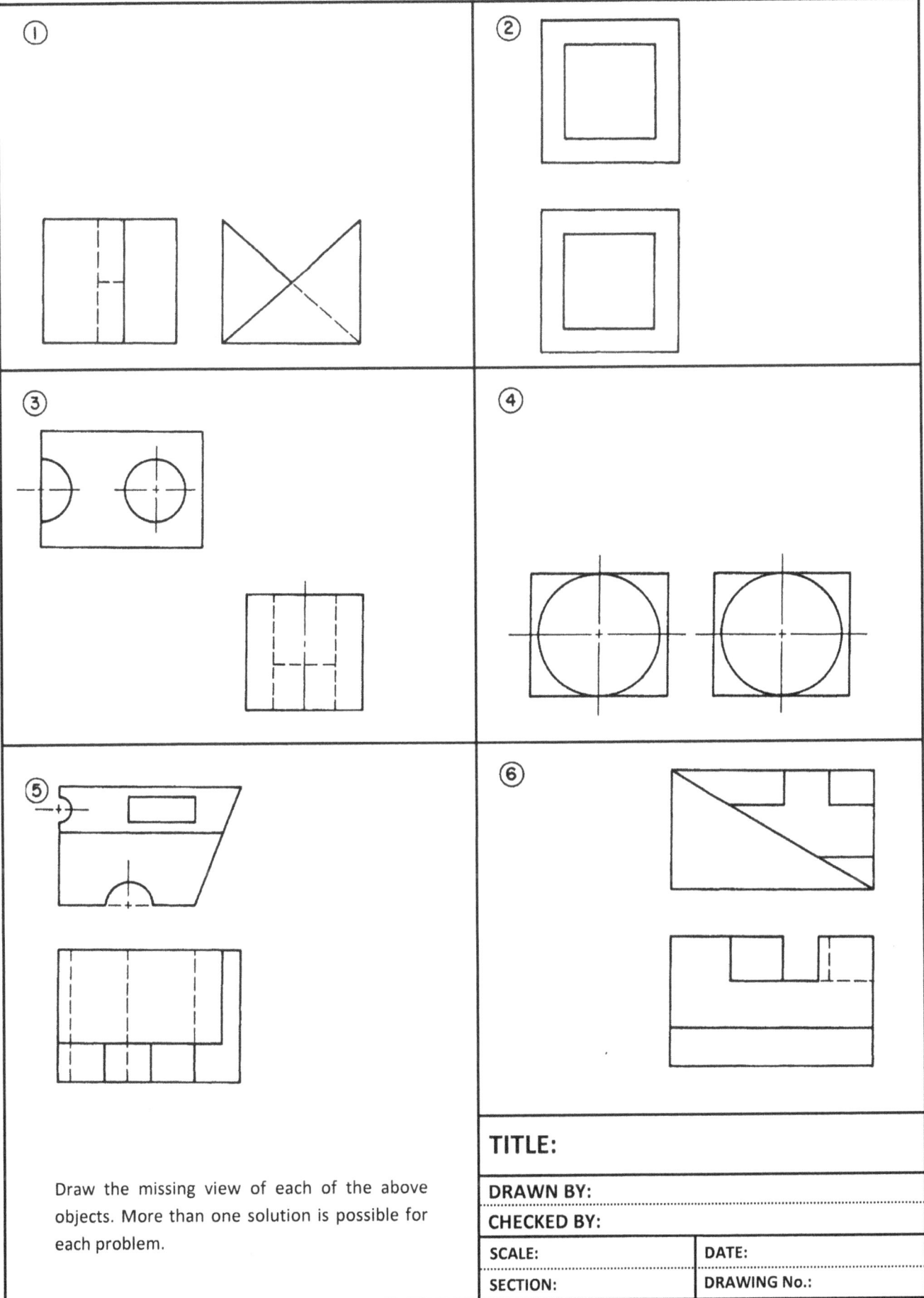

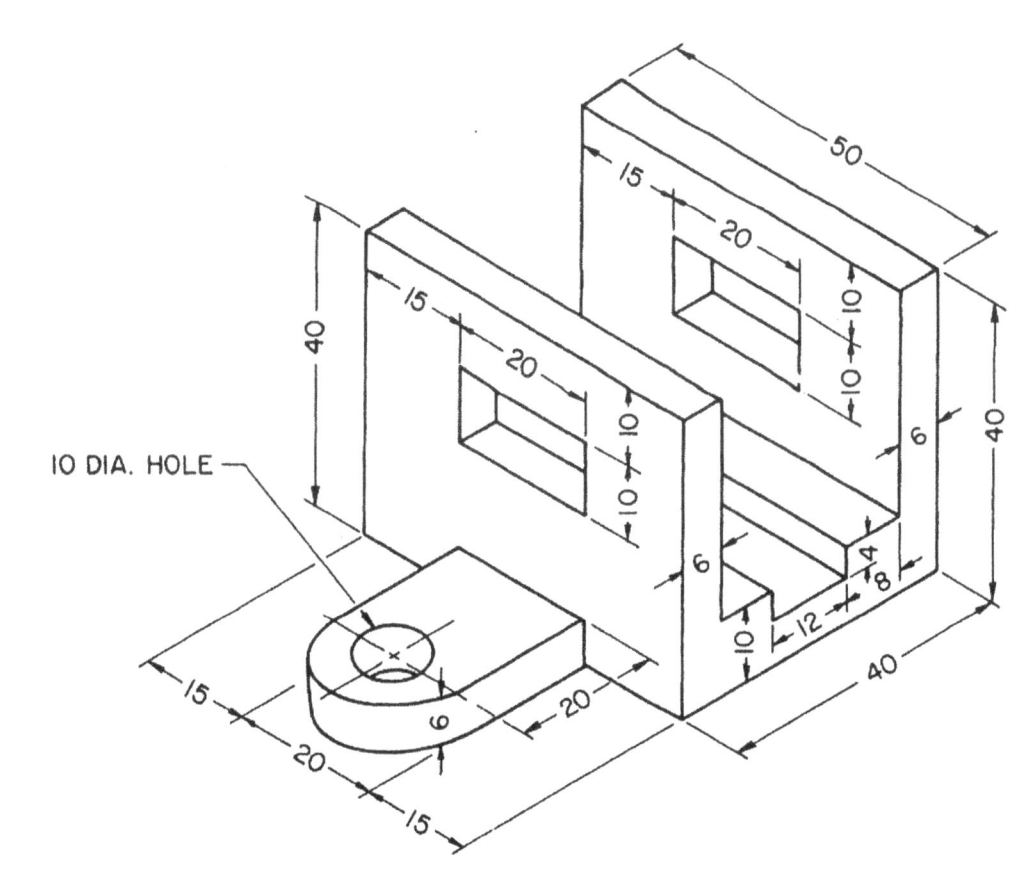

Draw, on the same sheet, three orthographic views to fully describe this object. Use scale 1:1. All dimensions are in millimeters.

TITLE:	
DRAWN BY:	
CHECKED BY:	
SCALE:	DATE:
SECTION:	DRAWING No.:

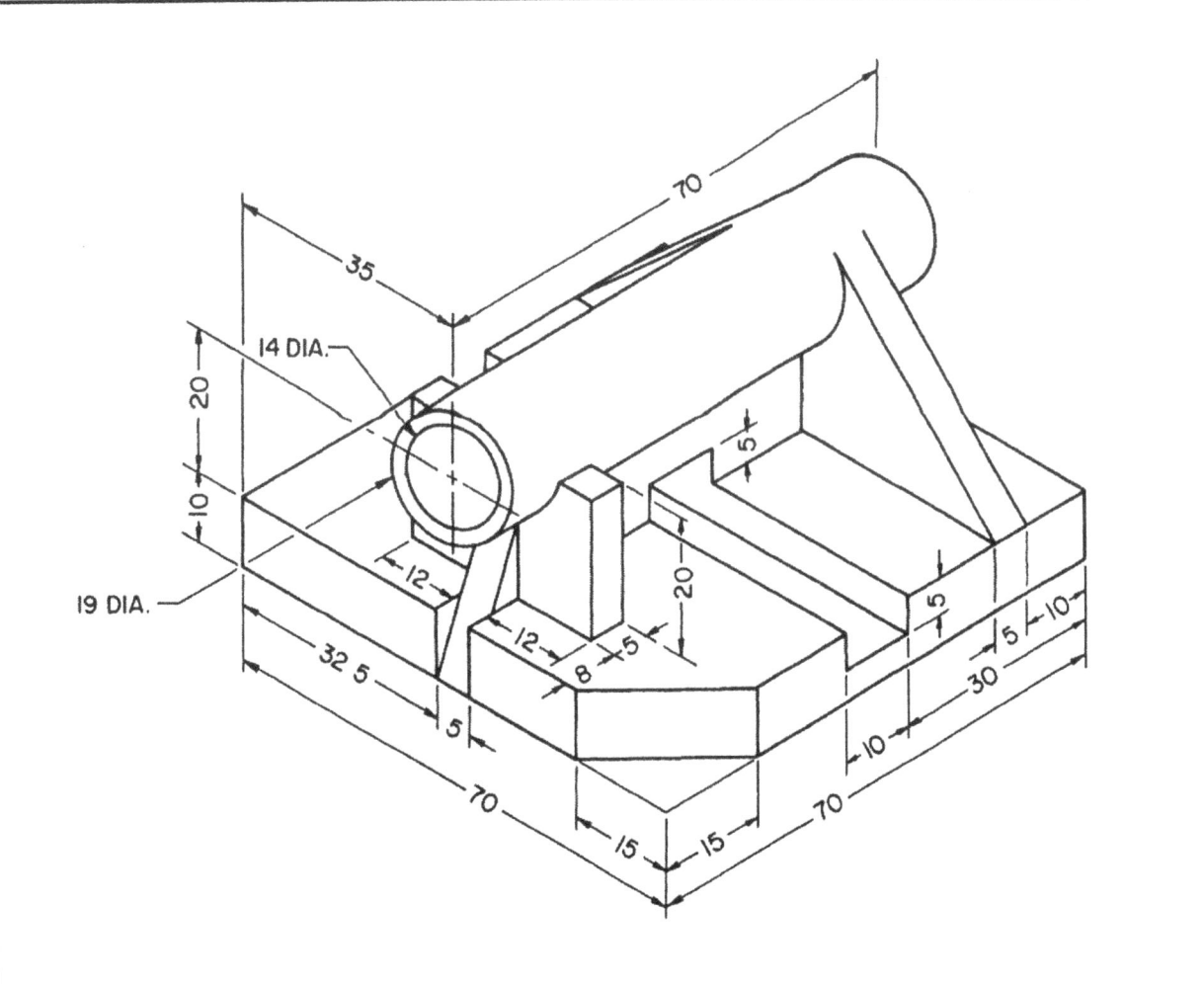

Draw the necessary orthographic views to represent the pipe sleeper shown above. Use scale 1:1. All dimensions are in millimeters.

TITLE:	
DRAWN BY:	
CHECKED BY:	
SCALE:	DATE:
SECTION:	DRAWING No.:

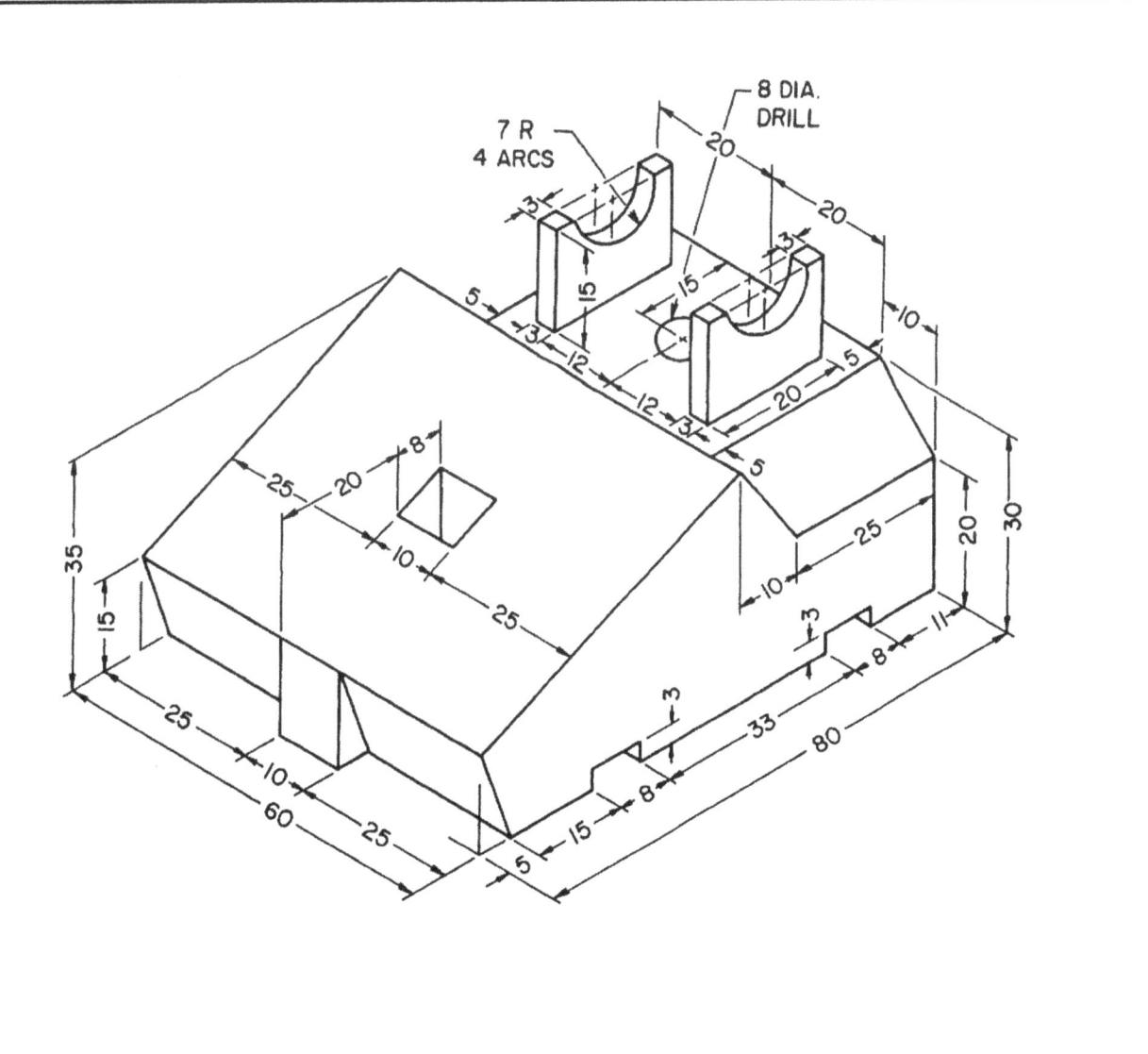

Draw the necessary orthographic views to fully describe the telephone shown above. Use scale 1:1. All dimensions are in millimeters.

TITLE:	
DRAWN BY:	
CHECKED BY:	
SCALE:	DATE:
SECTION:	DRAWING No.:

CHAPTER 5

AUXILIARY VIEWS

Views that are projected orthographically on planes other than principal ones are called "**Auxiliary Views**". The planes on which this kind of projection is made are called "**Auxiliary Planes**". These help in finding some information about the object which is not given by the principal views. One or more auxiliary planes can be used in the same problem.

Auxiliary views are used mainly to find the following:

- The true length of an oblique line
- The end view of a line
- The edge view of a plane surface
- The true shape of an oblique plane surface.

In order to find the above mentioned items, the auxiliary plane should be taken parallel to the line or plane under consideration if the true length of a line or the true shape of a plane is required. In the same sense, the auxiliary plane should be perpendicular to the line or plane under consideration if the end view of a line or the edge view of a plane is required. At the same time, in order to build a relationship between the auxiliary plane to be used and the principal planes already in use, the former should be constructed perpendicular to one of the latter ones. The line, common to both the auxiliary plane and the principal plane to which it is perpendicular, is called an "**Auxiliary Reference Line**". If a second auxiliary plane should be used, it could be perpendicular to the first auxiliary or to any one of the principal planes and so on and so forth.

In some cases, auxiliary views are used to help in the construction of principal views if these are hard to draw by ordinary orthographic projection methods. It is worth mentioning that, if auxiliary views are introduced in a certain problem, only the line or plane under consideration is projected and not the whole object unlike projections on principal planes.

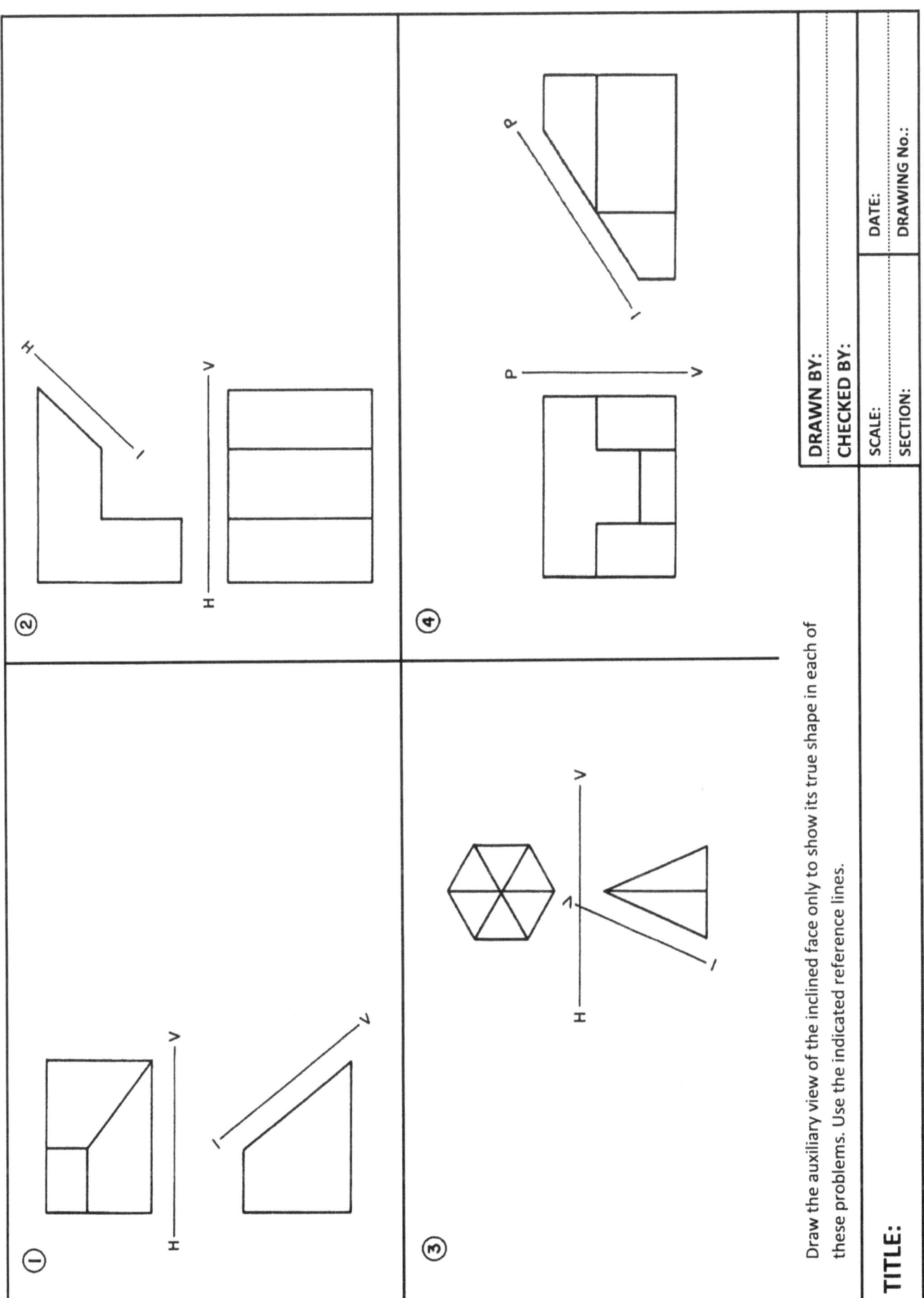

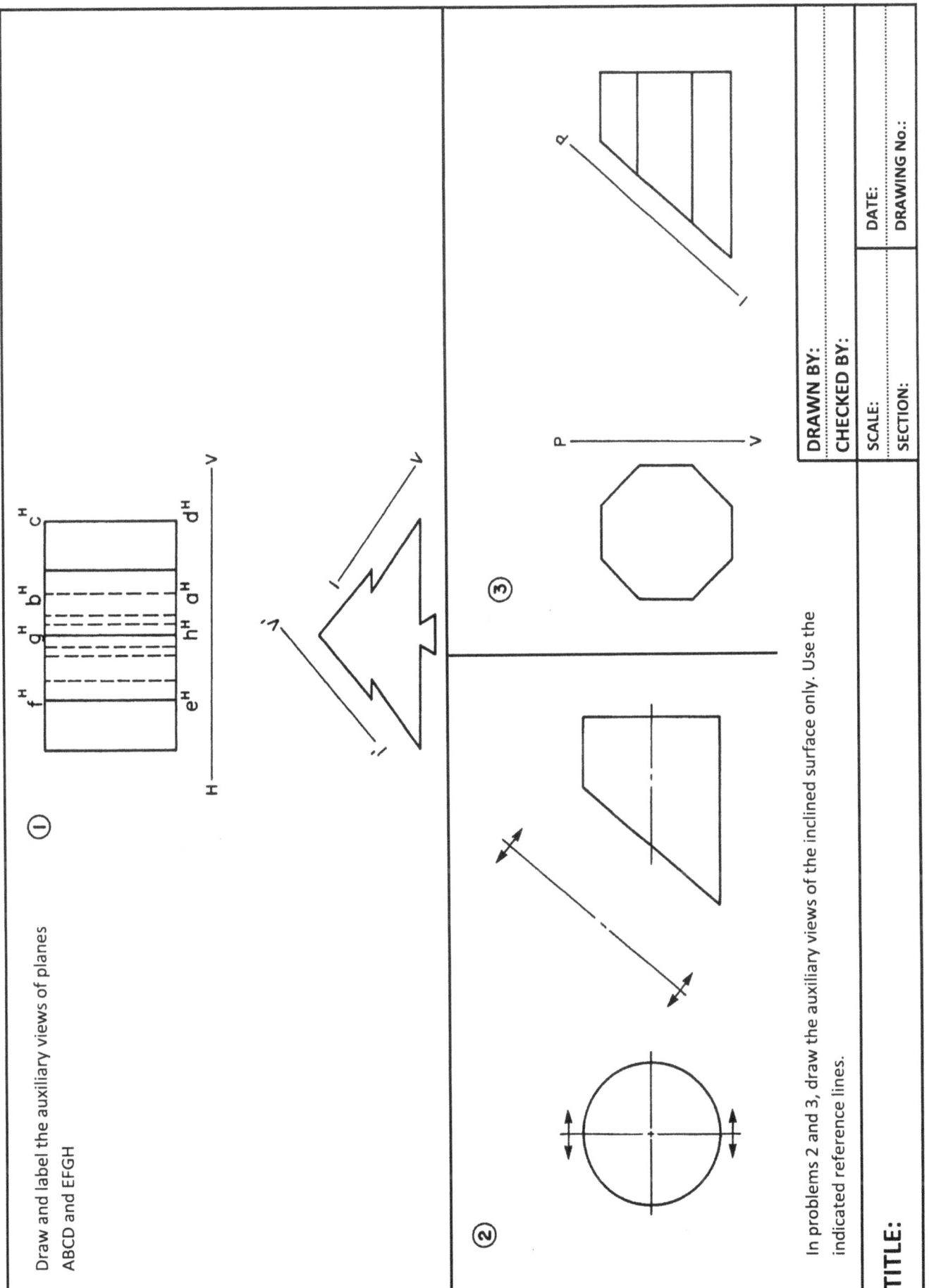

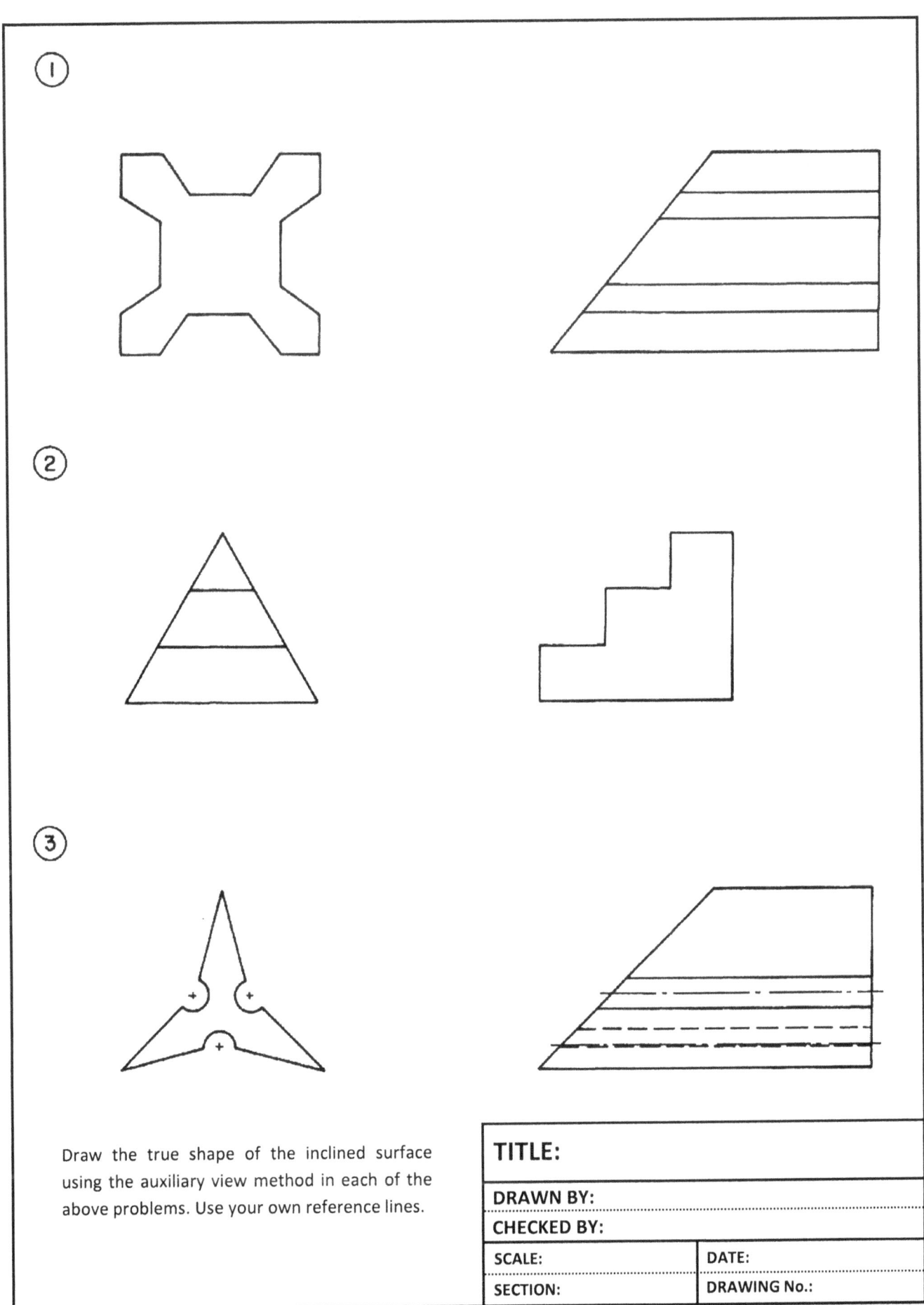

CHAPTER 6

SECTIONS

A **section** is a view obtained when the object is being imagined cut by means of a cutting plane and the portion closer to the observer being imagined removed to reveal the interior features of that object. Sections or sectional views will help in making the drawing easier to interpret since, without them, it will be crowded with many dashed lines representing the hidden features of a certain object.

In a sectional view, the solid parts of an object through which the cutting plane passes, through our imagination of course, are shaded with fine lines which are called section linings or crosshatchings. This is the conventional way of distinguishing a sectional view from a projected one. The conventional practice followed in crosshatching is as follows:

- Crosshatchings should always be inclined to the horizontal and 45° angle is generally used.
- They should be parallel to each other at a uniform space.
- They should be lighter than object lines. 4H pencil is generally used.
- They should be neither parallel nor perpendicular to any boundary line of the area being crosshatched.
- Use several slopes in crosshatching several adjacent parts of the same material.
- Crosshatch the same part in the same way even if it is at different places.
- Some prefer the use of different crosshatchings for different materials while others prefer the use of uniformly spaced lines and the material is specified by a note.

A sectional view is always accompanied with one or more projected views in order to complete the description of the object. A cutting plane line is drawn on the appropriate projected view to indicate where the cut is made and arrowheads are added to show the direction of viewing the sectioned portion.

In a sectional view, all visible lines beyond the cutting plane are drawn whereas all hidden lines are omitted unless some of them are needed for dimensioning or to make the drawing clearer.

Types of Sections
The types of sections that are used in engineering drawings include: Full, Half, Offset, Aligned, Revolved, Removed, Auxiliary and Broken Out sections.

Conventions
In sectional views many conventions are used and some rules of projection are violated in order to make the drawing easier to read and to avoid misunderstanding. As a matter of fact, it is said in these respects that if any clarity or understanding could be achieved by violating a rule, violate it. Some of the conventions that are commonly used are the following:

- If the cutting plane passes through the longitudinal axes of solid cylinders such as shafts, bars, screws, bolts, etc., these parts are not crosshatched as if the cutting plane didn't pass through them because there is no interior features to disclose in such items.
- If the cutting plane passes through the longitudinal axes of spokes of a wheel and thin webs, these are not crosshatched in order to distinguish them from each other and from different parts of the object. Whereas, if the cutting plane is perpendicular to the axis of the web or spoke, then a cross section (revolved section) is obtained and this should be crosshatched.
- In a symmetrical object, if there is an odd number of axes of symmetry such as holes, spokes or webs, and the cutting plane passes through one of these items, the others are considered as rotated around until they lie on the section plane and then the view is projected. This is done in both sectional and projected views otherwise the object does not seem symmetrical and the views will be crowded with many hidden lines.

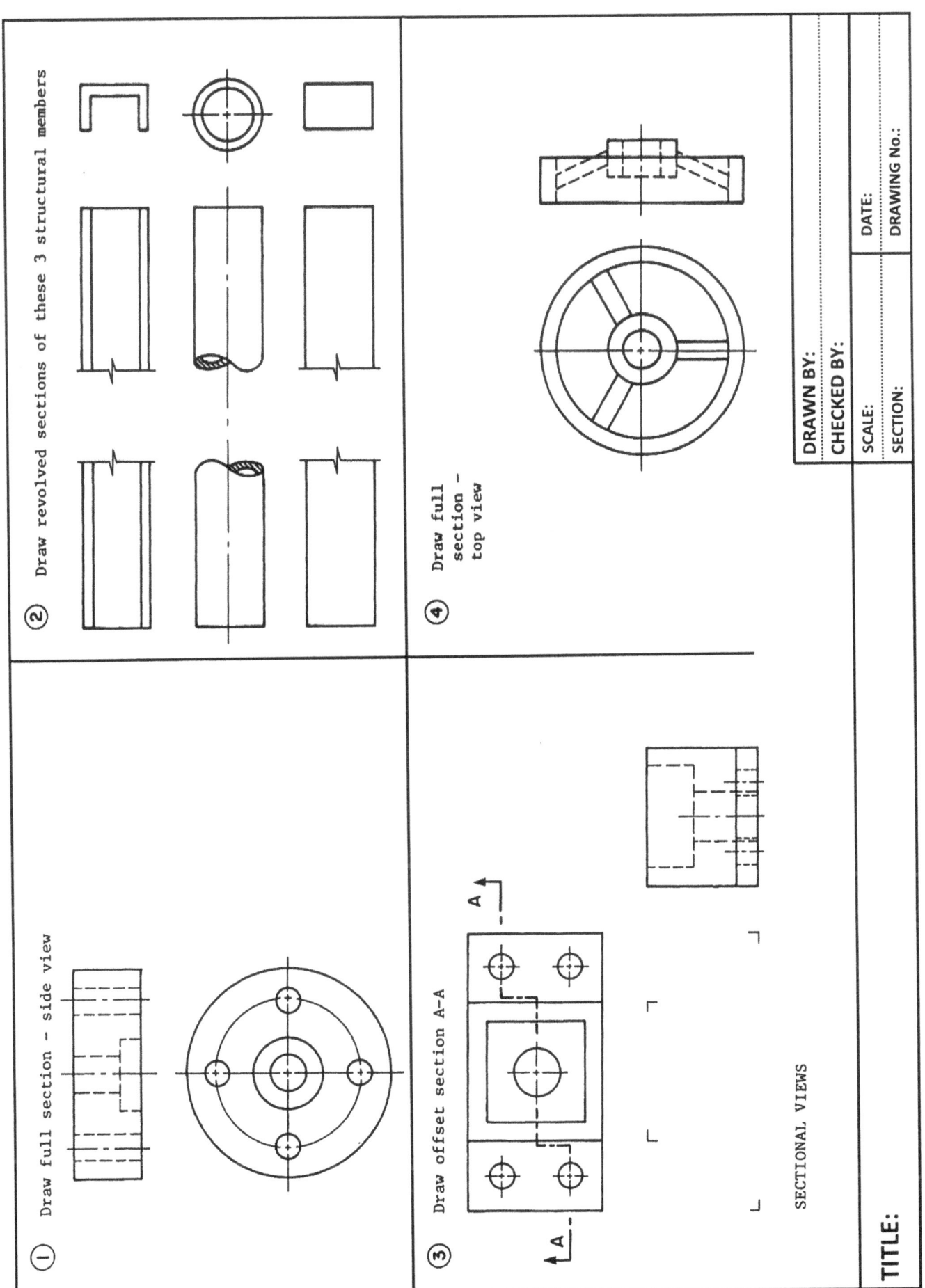

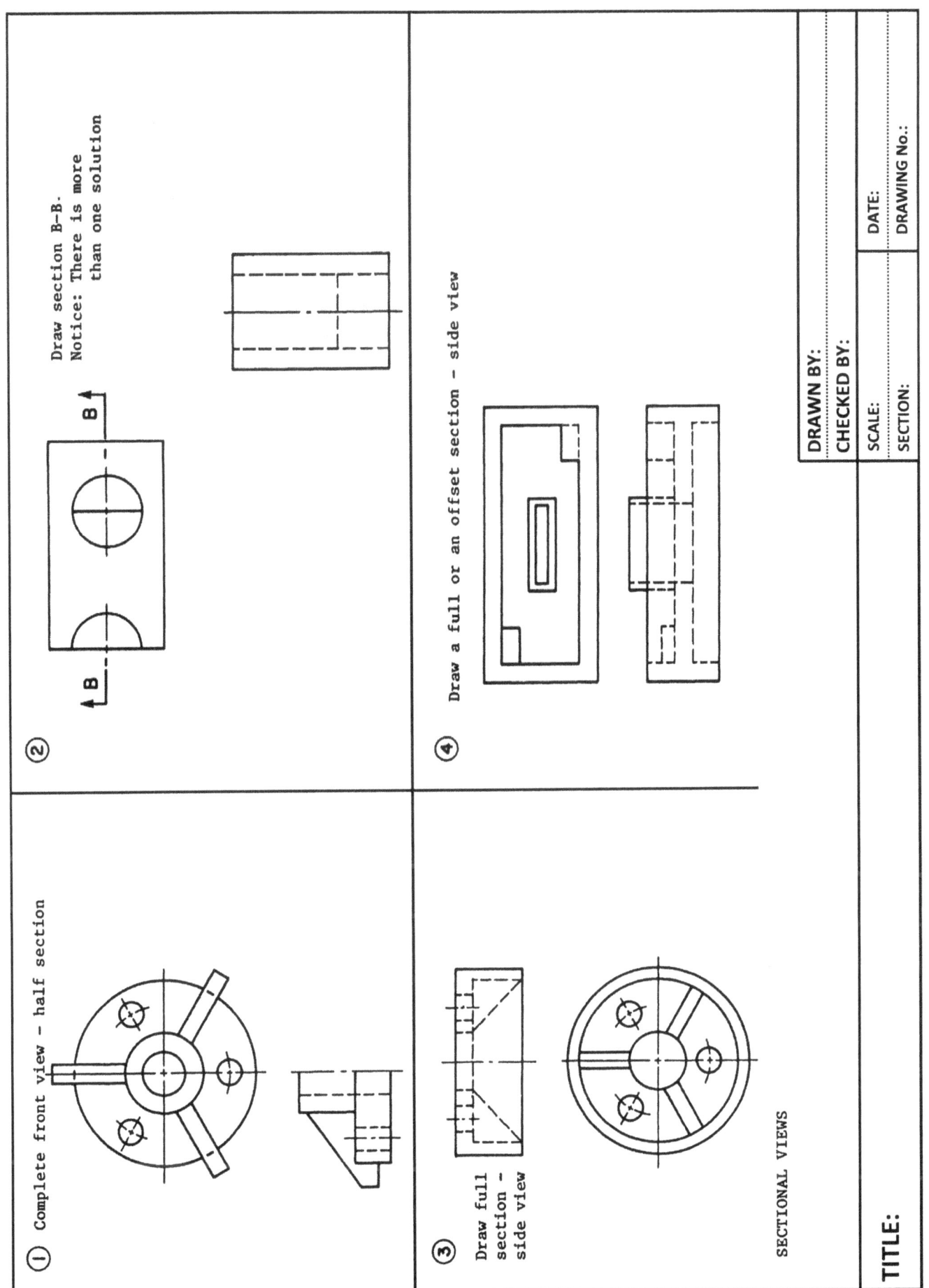

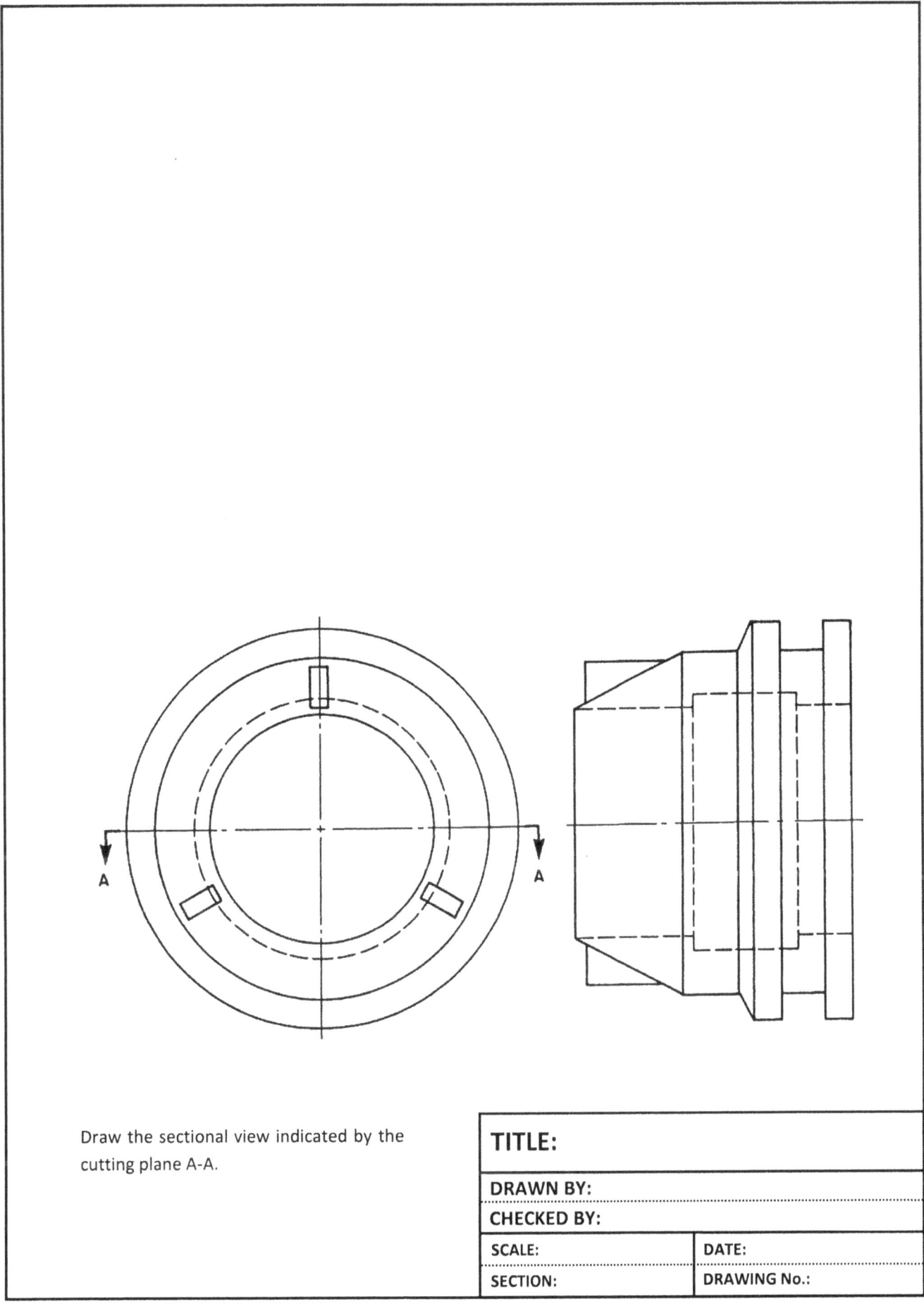

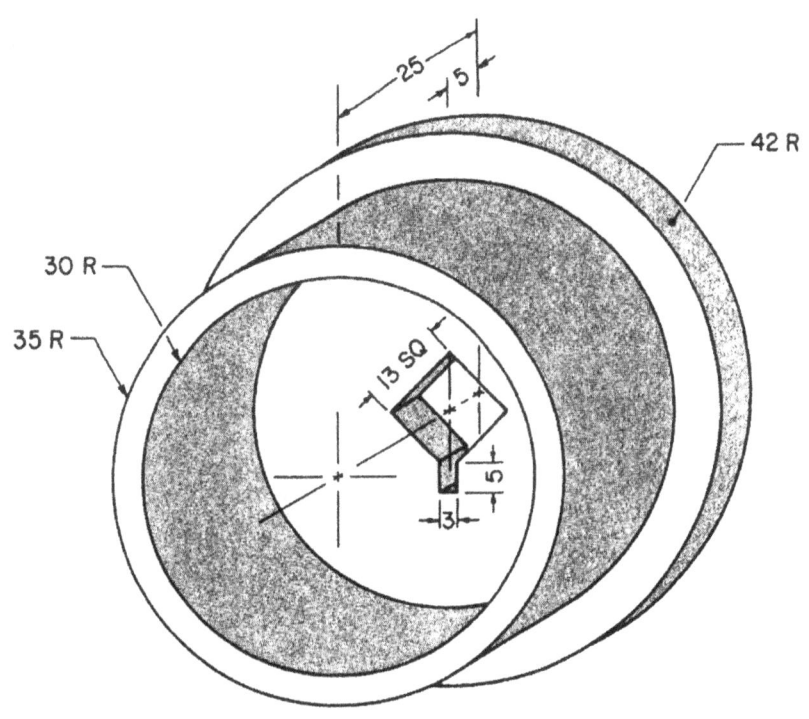

Draw the necessary orthographic and sectional views to fully describe the flow regulator shown in Cavalier drawing. Use scale 1:1. All dimensions are in millimeters.

TITLE:	
DRAWN BY:	
CHECKED BY:	
SCALE:	**DATE:**
SECTION:	**DRAWING No.:**

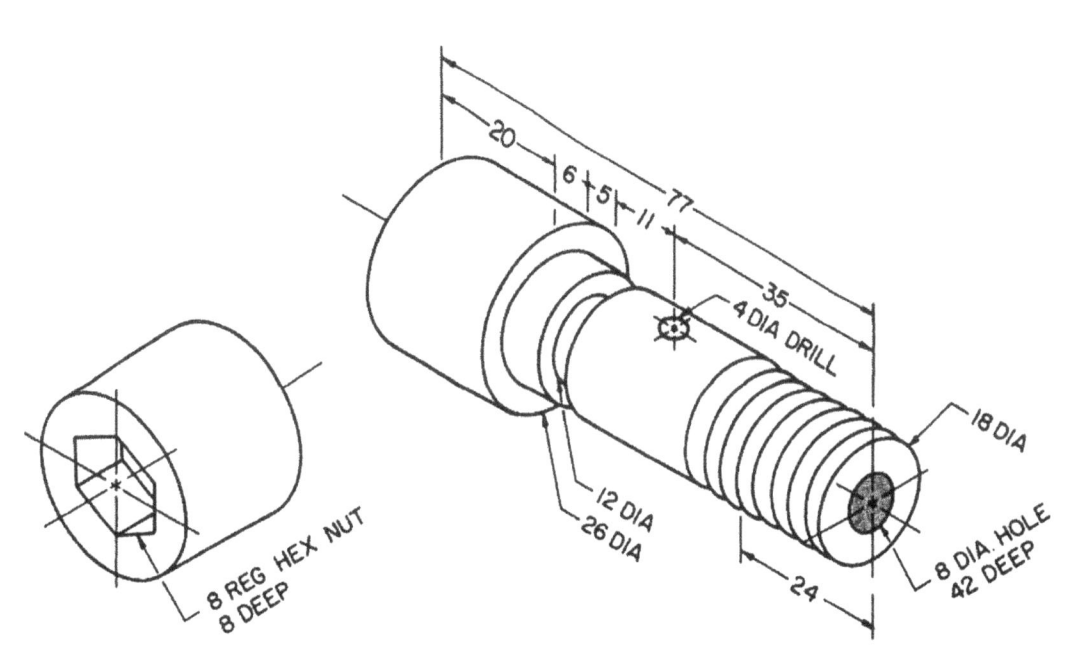

Draw the necessary orthographic and sectional views to fully describe the special bolt shown in isometric. Use scale 1:1. All dimensions are in mm.

TITLE:	
DRAWN BY:	
CHECKED BY:	
SCALE:	DATE:
SECTION:	DRAWING No.:

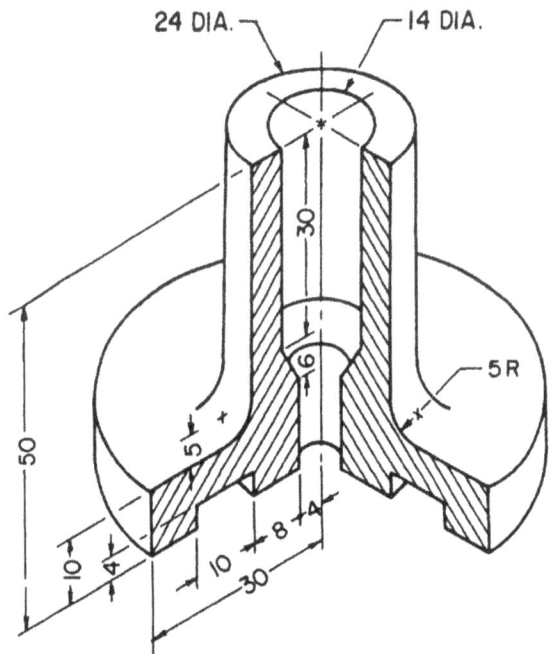

Draw the necessary orthographic and sectional views to fully describe the special valve shown in isometric. Use scale 1:1. All dimensions are in mm.

TITLE:	
DRAWN BY:	
CHECKED BY:	
SCALE:	DATE:
SECTION:	DRAWING No.:

CHAPTER 7

DIMENSIONING

A drawing is not complete unless dimensions and necessary notes are added. Although a drawing is made to a certain scale, it is not permitted in any circumstances to scale out dimensions in order to construct that object or any part of it. In lieu, only the written dimensions are used. In order to add dimensions that are clear and can be easily read, certain rules should be followed. These could be divided into two categories: The first one governs the technicality of drawing the lines that make up the dimensioning system and the second governs positioning and locating dimensions on a drawing.

I. Rules of Technique (Technical Rules)

- Extension line is a thin full line.
- Extension line extends to within 1.5 mm (1/16 inch) of object.
- Extension line extends about 3 mm (1/8 inch) beyond last arrowhead.
- Dimension line is a thin full line.
- Dimension line is broken at point of dimension.
- Closest dimension line is 8 mm (5/16 inch) from object.
- Dimension lines should be 8 mm (5/16 inch) apart.
- Dimension lines are perpendicular to extension lines.
- Arrowheads are long and narrow, about three times as long as wide.
- Arrowheads must just touch extension lines.
- Arrowheads must be very dark.
- A leader is a thin, straight, full line terminated by an arrowhead if it points to a line, curved or straight, and by a dot if it points to an area.
- A short horizontal line (4-5 mm long) should be added at the other end of the leader at the level where the note is to be written.
- Letters and numerals should be large enough to be easily read. They should be at least 3 mm (1/8 inch) high.
- Capital letters are preferred.
- Fractions should be 1 ½ times the height of the integer.
- Fraction dividing line should be in alignment with the dimension line.

II. Rules for Locating Dimensions

- Dimensions should read from left to right and from bottom to top of a drawing; diameters and radii are excepted. Recently, a horizontal unidirectional dimensioning system is generally implemented.

- Dimensions are placed outside the view unless putting them that way results in long extension lines.
- Dimensions should be placed between the views, but closest to the appropriate view.
- Dimensions should be placed on the view that is most descriptive.
- Dimensions should be placed approximately at the center of the dimension lines whereas adjacent dimensions should be staggered to avoid crowding.
- Avoid cumulative tolerances by referring the dimensions to reference lines.
- In referring dimensions to a reference line, place the smallest dimension closest to the view in order to avoid extension line crossings.
- Detail dimensions in the same series should be lined up with each other. An overall dimension must be added.
- Do not dimension any line twice (i.e., do not repeat dimensions).
- Do not dimension to an invisible line.
- Do not cross dimension lines.
- Avoid superfluous dimensions.
- Dimension to center of circles and not to their edges.
- Hole patterns should be dimensioned on the view where they are shown as circles.
- Cylinders are dimensioned on the view where they are shown as rectangles.
- Sizes of circles are usually given by leaders which should touch the circle or arc pointing towards their centers.
- Avoid leaders crossing each other, long leaders, leaders in horizontal and vertical positions, and leaders parallel to adjacent dimension lines.
- Show finish marks on all surfaces to be machined (or polished).
- Sizes and location of all features should be given.
- Do not force worker to calculate to obtain dimensions.

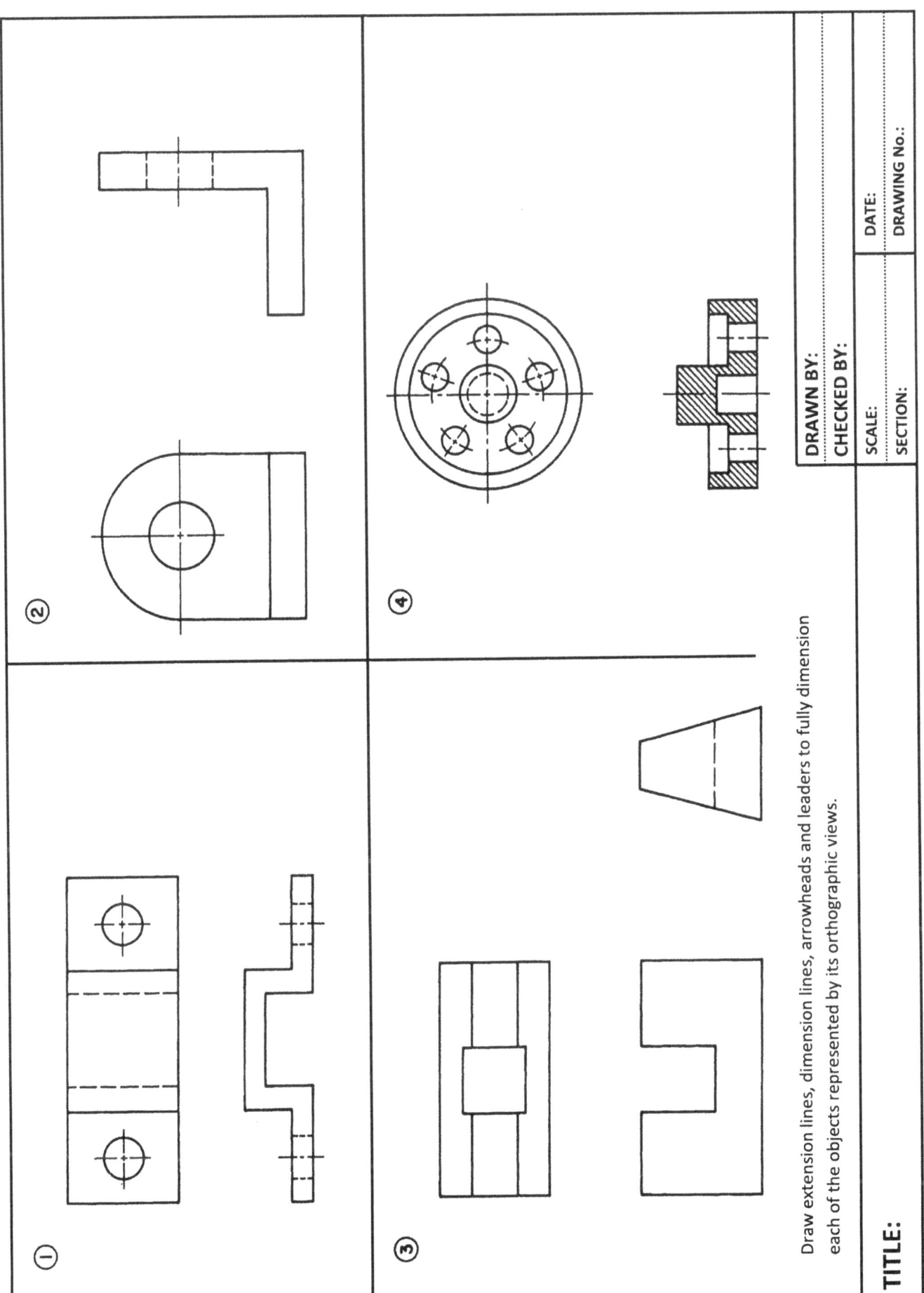

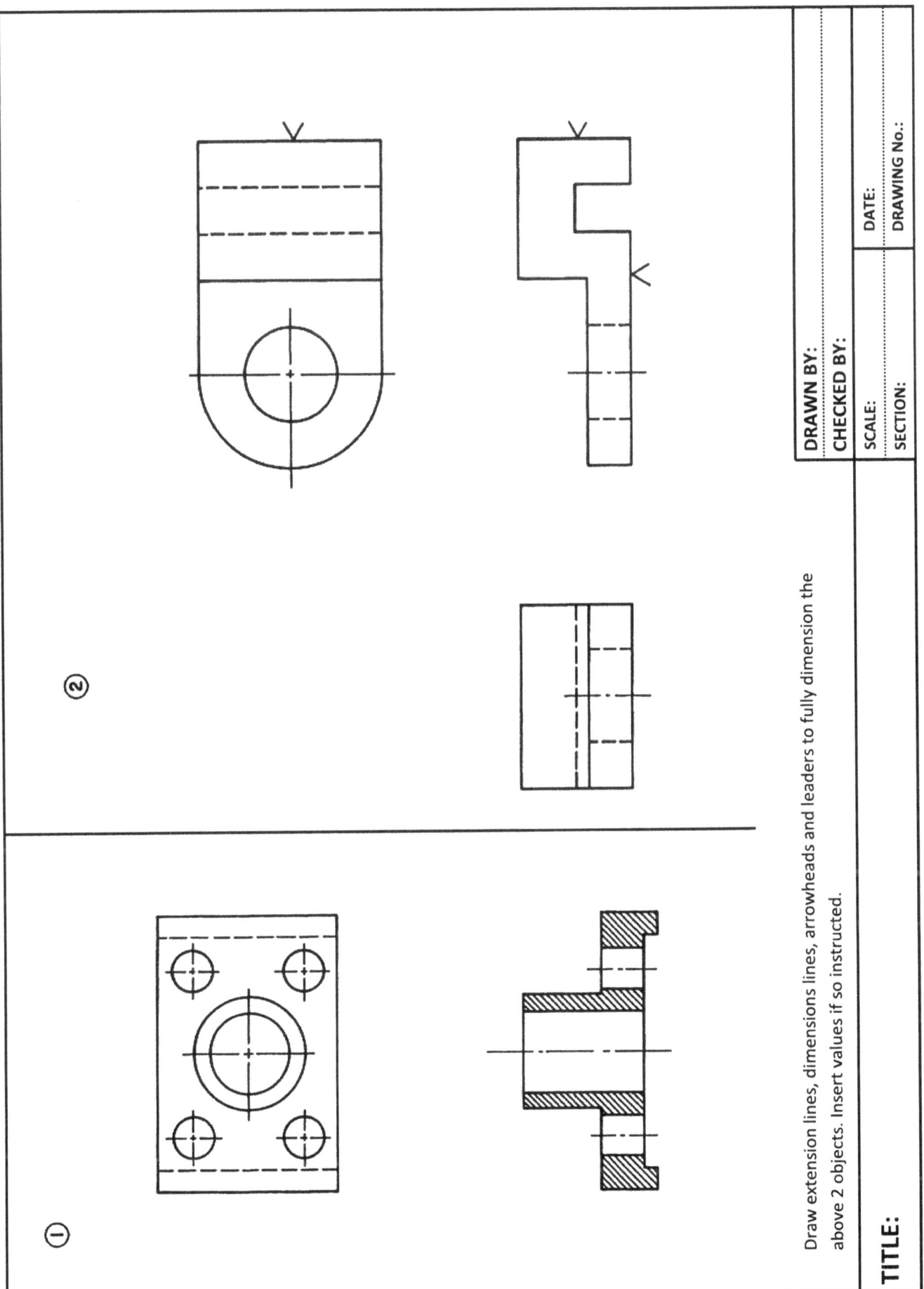

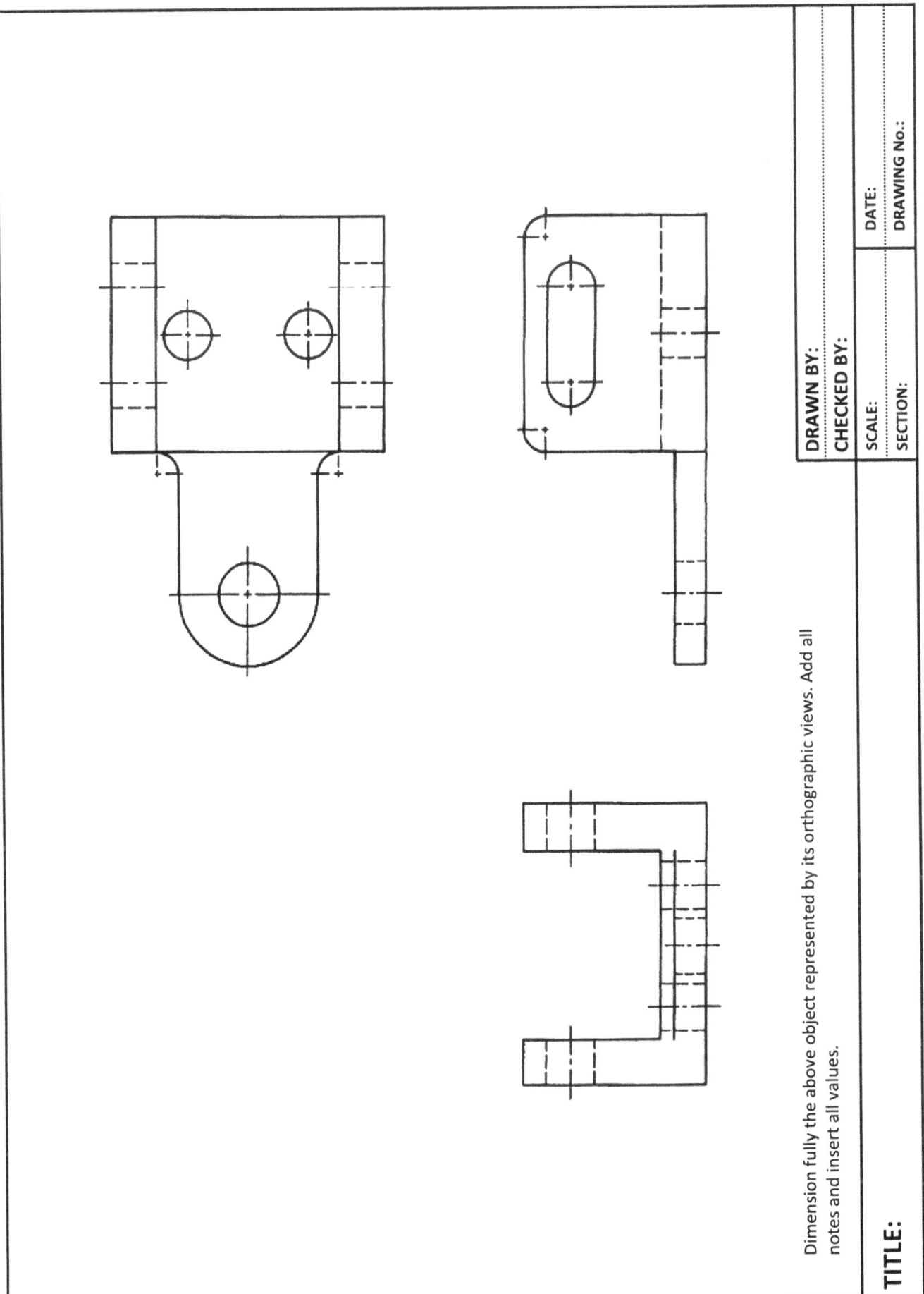

Dimension fully the above object represented by its orthographic views. Add all notes and insert all values.

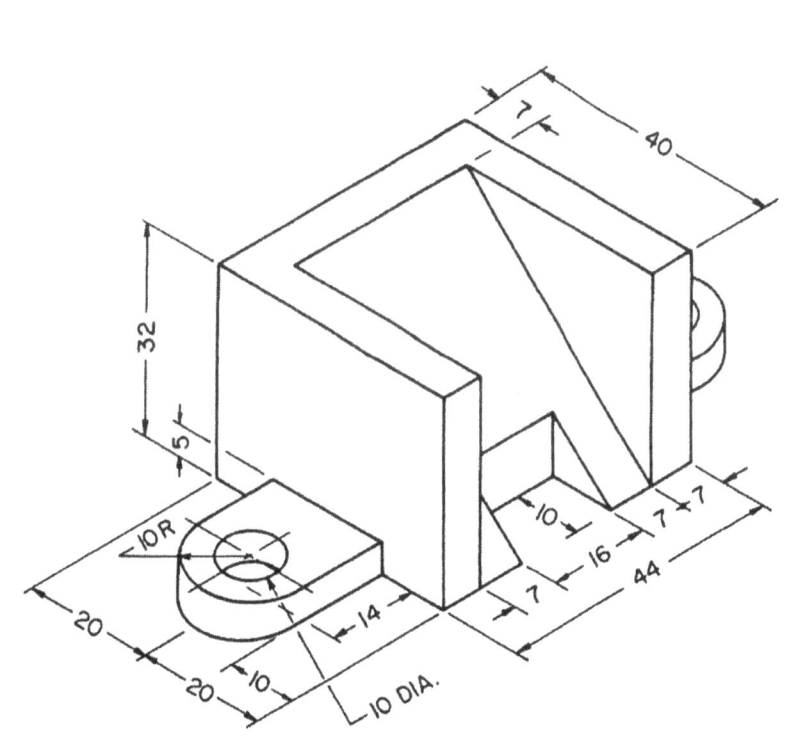

Draw the necessary orthographic views to fully describe this object. Dimension if so instructed. Use full size scale. All dimensions are in mm.

TITLE:	
DRAWN BY:	
CHECKED BY:	
SCALE:	DATE:
SECTION:	DRAWING No.:

CHAPTER 8

PICTORIAL DRAWINGS

A drawing which shows three faces of the object at the same time is called a **"Pictorial Drawing"**. It enables one to see an object in its three dimensional form. It employs one plane of projection which is called the **"Picture Plane"**. There are three major kinds of pictorial drawings:

- Axonometric
- Oblique
- Perspective.

I. Axonometric Projection

Theoretically, the faces of the object, in this projection, are inclined to the picture plane, whereas the lines of sight, which are parallel to each other, are perpendicular to it. Practically, the object is drawn within a rectangular box whose axes are parallel to the three major dimensions of the object. Two of these axes are inclined to the horizontal and one is vertical. There are three forms of axonometric projections or drawings which are widely used. These are:

- **Isometric**
- **Dimetric**
- **Trimetric.**

In an **isometric** drawing, the horizontal axes of the cube are usually drawn at an angle of 30° with the horizontal and the vertical one remains vertical. In other words, these axes make an angle of 120° with each other. It is worth mentioning that the same scale is used on all three axes. A circle in an isometric drawing appears as an ellipse irrespective of the face it lies in. This ellipse could be drawn exactly by the "Coordinate Method" and approximately by the "Four-Center Approximate Method".

In order to construct an isometric drawing of a certain object, one should have the orthographic views to start with. Then, the following steps are followed:

- Draw the orthographic views to the same scale as that to be used on the isometric drawing.
- Draw the smallest rectangular box that can circumscribe that object in its isometric form.
- Draw the different parts of the object in that box in their respective isometric planes.
- Omit all invisible lines. It is a usual practice to omit these lines in all pictorial views.

A property of an isometric drawing is that it can be dimensioned. All of the dimensioning rules studied so far apply here except for the fact that dimension and extension lines should be made to lie in isometric planes. An isometric plane is any one which is formed by two axes.

In a **dimetric** drawing, the same scale is used on any two axes and a different one on the third. In a **trimetric** drawing, three different scales are used, one on each of the three axes. The method of construction of a dimetric and a trimetric drawing is the same as that used in isometric drawings except that the orthographic views are drawn to the scales used in either the dimetric or the trimetric drawings.

II. Oblique Projection

In this kind of pictorial drawing, one plane of projection is used as in the case of the axonometric projection, but the lines of sight, although parallel to each other, are inclined (oblique) to the picture plane. The object here could be placed in any position, but for convenience, it is common practice to place the front face of the object parallel to the picture plane.

There are three main types of oblique projections which are:

- **Cavalier**
- **Cabinet**
- **General Oblique.**

In constructing an oblique projection, the box method is used as in the case of an axonometric projection. But, the front face axes are perpendicular to each other, one is horizontal and the other is vertical. The third which is called the "Receding Axis", could be at any angle to the horizontal. The different types of oblique projections differ from each other; theoretically, by the angle the lines of sight make with the picture plane, and practically, by the scale used on the receding axis. The front face axes have the same scale in all types.

In the practical method of constructing an oblique projection, the same scale is used on all axes in **cavalier** drawings. In **cabinet** drawings, the scale used on the receding axis is equal to half of that used on the front face axes. In general oblique drawings, the scale used on the receding axis could be any ratio to that used on the front face axes except one and one half.

Since the front face of the object is parallel to the picture plane, it is advisable to make the face with many circles as the front one because these will appear as circles and not as ellipses in that face and it is easier to draw a circle than to draw an ellipse. As a matter of fact, the front face of an oblique projection appears as its orthographic view. As is the case in **axonometric** drawings,

an oblique drawing can be dimensioned. All of the dimensioning rules apply here except that dimension and extension lines should lie in the same oblique plane.

III. Perspective

In this form of pictorial drawings, the lines of sight are no more parallel but they converge to a point which is the point of sight or the eye of the observer. Because of this fact, the perspective is the only form which most nearly approaches the picture as seen by the eye. As is the case with the other pictorial views, one picture plane is used whose position changes with the change of that of the point of sight. That is why one can have a perspective by looking through a horizontal or a vertical plane. The latter is the one mainly used in the housing industry.

Many methods are used in constructing a perspective which could be found in engineering graphics books or in an architecture book dealing with this subject. A perspective cannot be dimensioned. This makes its usage qualitative. For this reason, perspectives are mainly used in the housing and building projects where the photographic shape of the house or building is shown before they are constructed.

①

②

Draw, in the space to the right, an isometric drawing of each of the two objects.

TITLE:	
DRAWN BY:	
CHECKED BY:	
SCALE:	DATE:
SECTION:	DRAWING No.:

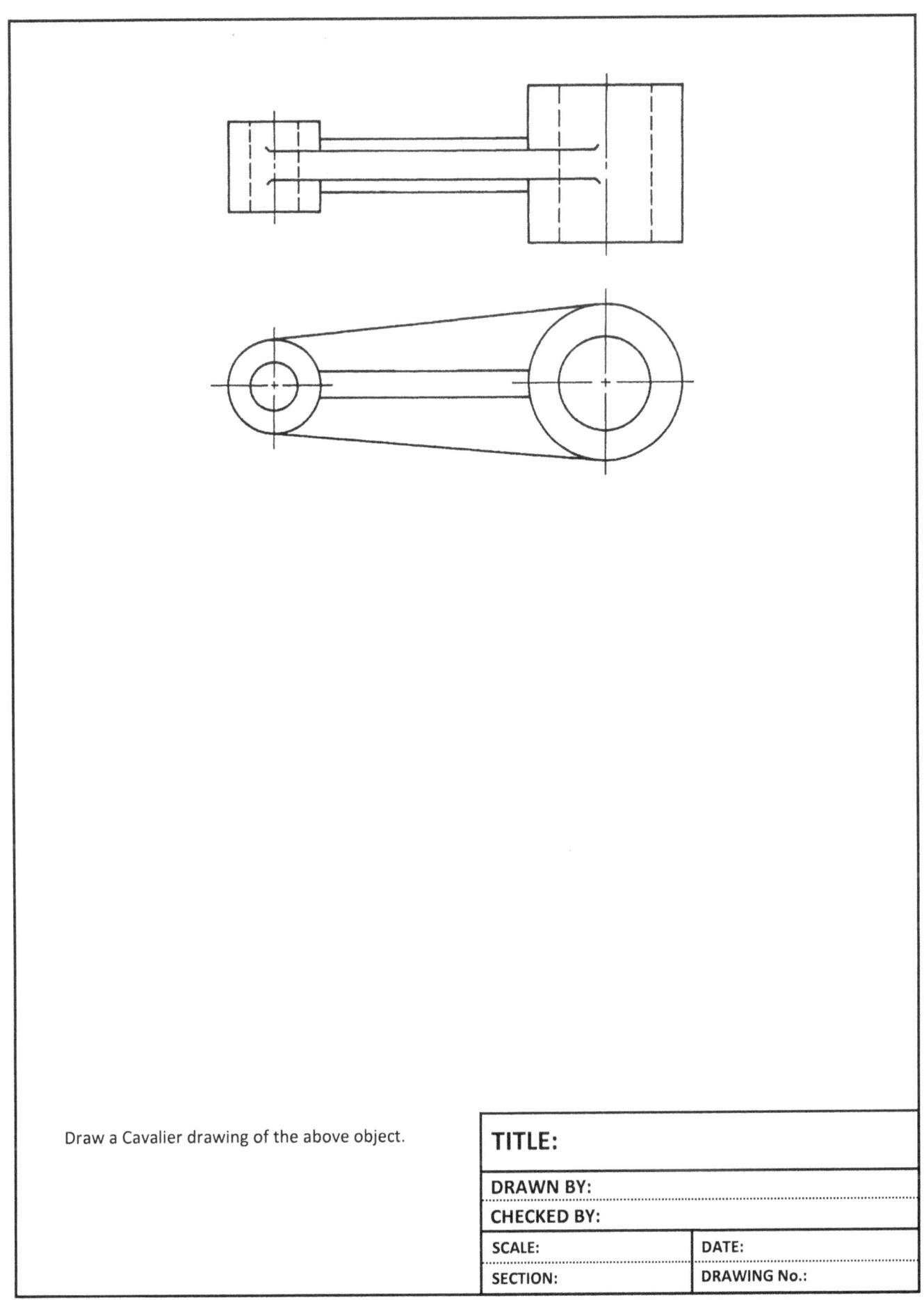

Draw a Cavalier drawing of the above object.

TITLE:	
DRAWN BY:	
CHECKED BY:	
SCALE:	DATE:
SECTION:	DRAWING No.:

Draw a dimetric or a Cabinet drawing of the above object. The scale is to be given by your instructor.

TITLE:

DRAWN BY:

CHECKED BY:

SCALE:

DATE:

SECTION:

DRAWING No.:

Draw an isometric or a Cavalier drawing of the above object as instructed by your instructor.

TITLE:	
DRAWN BY:	
CHECKED BY:	
SCALE:	DATE:
SECTION:	DRAWING No.:

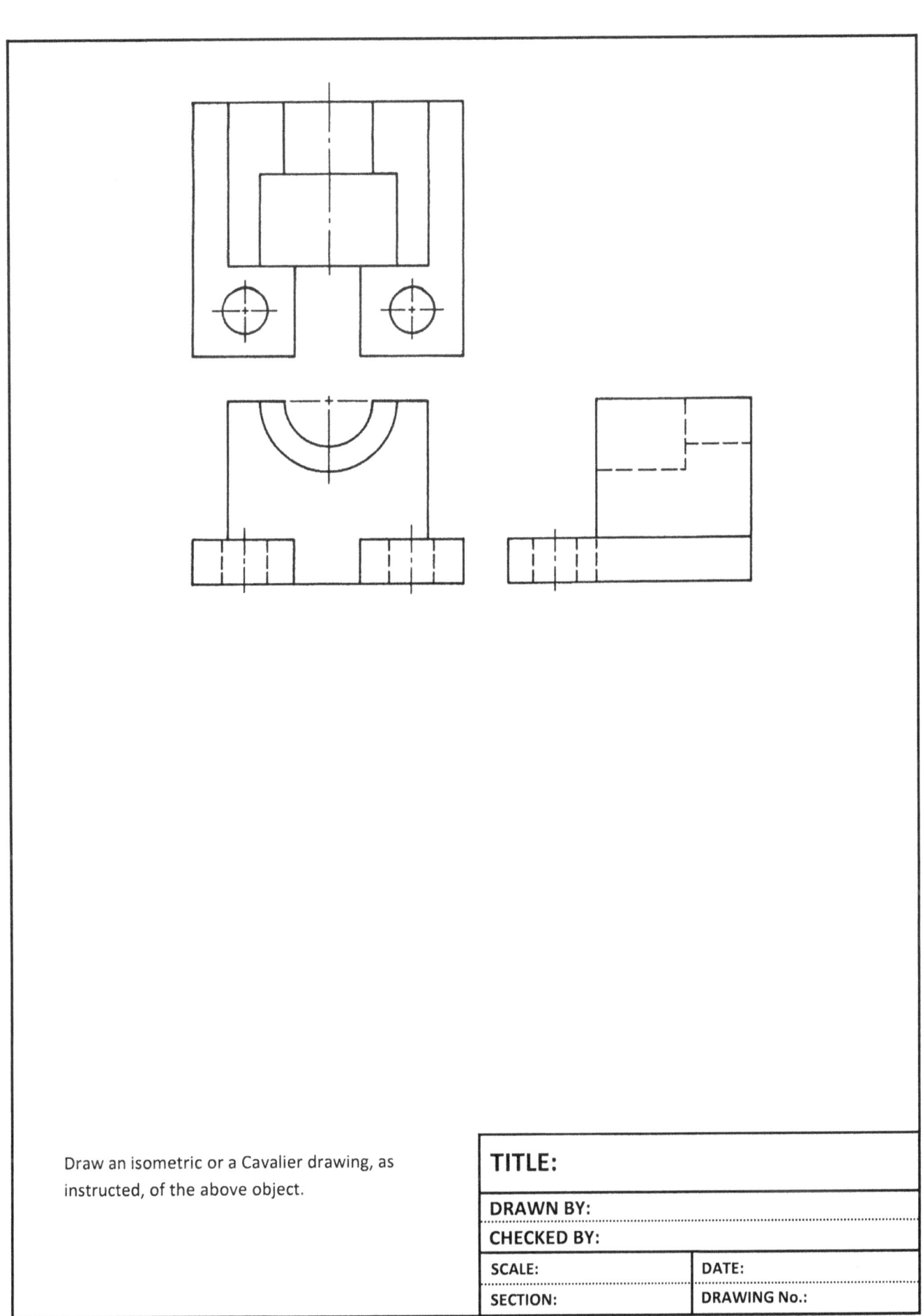

Draw an isometric or a Cavalier drawing, as instructed, of the above object.

TITLE:	
DRAWN BY:	
CHECKED BY:	
SCALE:	DATE:
SECTION:	DRAWING No.:

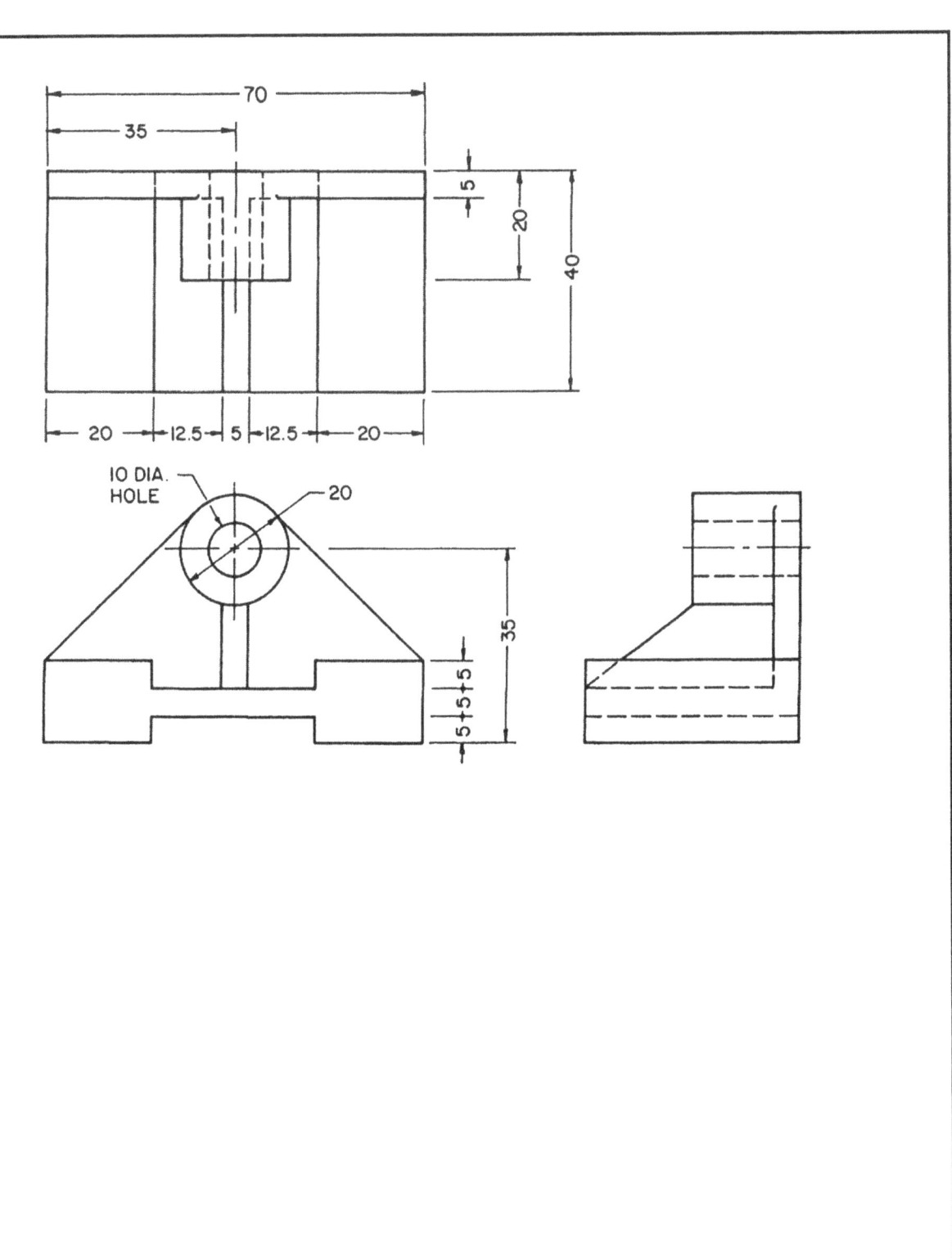

Draw and dimension an isometric or a Cavalier drawing of the above object. All dimensions are in millimeters.

TITLE:	
DRAWN BY:	
CHECKED BY:	
SCALE:	DATE:
SECTION:	DRAWING No.:

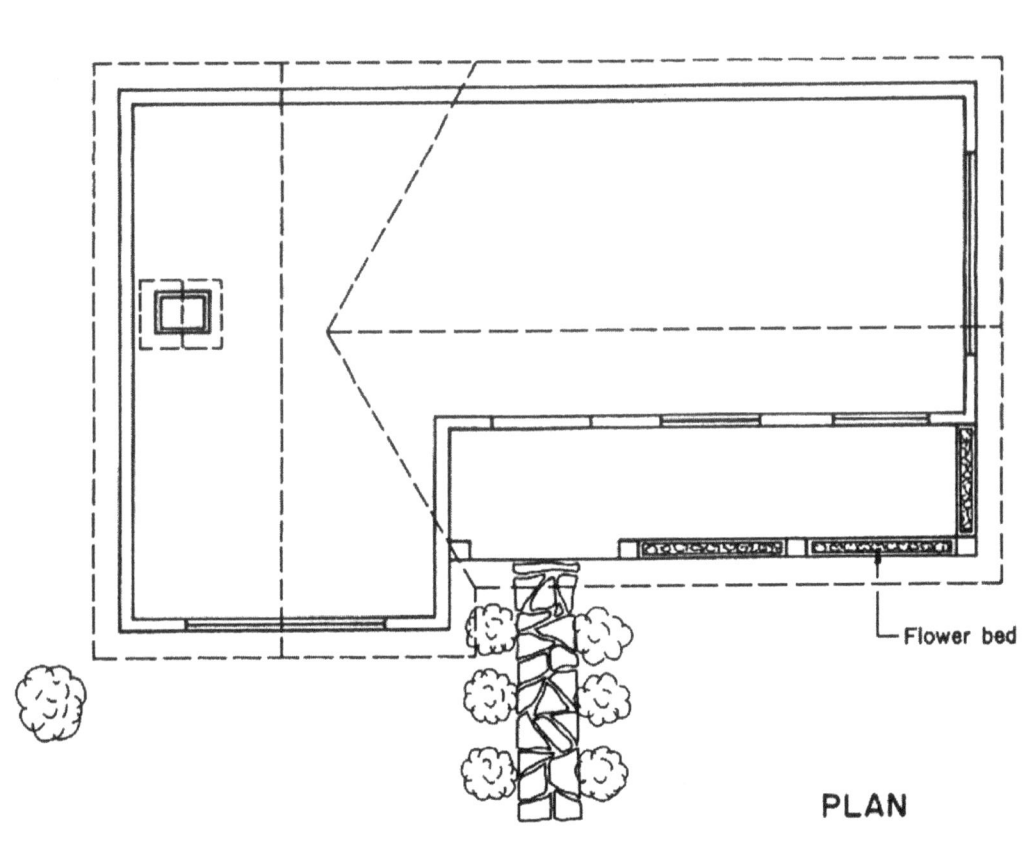

PLAN

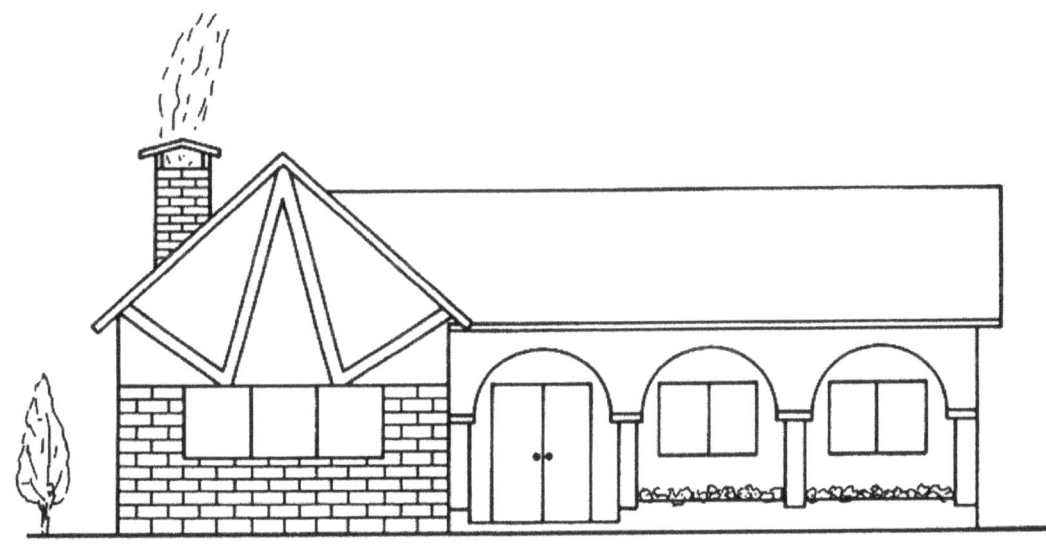

ELEVATION

Draw, on a separate sheet, an angular or two-point perspective of the above house. Use your imagination to do the landscaping or to treat the exterior differently. The scale of the plan and elevation is 1:100.

TITLE:	
DRAWN BY:	
CHECKED BY:	
SCALE:	DATE:
SECTION:	DRAWING No.:

CHAPTER 9

FASTENERS

Fasteners are devices that are used to hold the different parts of a machine or a structure together. They vary from simple nails and screws to rivets and heavy bolts. These days, welding is used on a large scale to form permanent connection. Some of these fasteners, such as nails and small screws, are not represented on a drawing and some others, such as bolts and rivets, are represented by conventional symbols.

Fasteners such as standard bolts and screws use threads whose basic curve is the "Helix". This can be generated by a point moving on the circumference of a cylindrical or conical surface at a constant angular speed with a simultaneous uniform rate of advance in an axial direction. These threads could be right-handed or left-handed. A right-handed thread is that which advances away from the observer if turned clockwise, whereas a left-handed thread advances away from the observer if turned counter-clockwise.

The various types of threads found in the market are distinguished from each other by the shape of their profile. These are classified as follows: **Sharp V, American National, Whitworth, and the knuckle threads**. These are general purpose threads. Then, there are the **Square (or the modified square), the Acme, the Buttress and the Worm threads** which are used to transmit power. In order to find how these threads are represented on a drawing, refer to any engineering graphics textbook.

These threads are also divided into different series based on the distance between two consecutive threads which is called the pitch which is normally given in the form of number of threads per inch. The following thread series are used according to the American standards:

1. Unified National Coarse series (UNC)
2. Unified National Fine series (UNF)
3. Unified National Extra Fine series (UNEF)
4. Unified National Miniature series (UNM)
5. Unified Pitch series (UN)

Besides representing a thread by a symbol on a drawing, it should be accompanied by a specification note. For example: 5/8 - 9 UNC - 2A; where 5/8 is the nominal (major) diameter in inches, 9 is the number of threads per inch, UNC is the thread series, 2 is the class off it and A is the letter used to represent external threads (the letter B is used for internal threads). The class of fit represents the ease with which two mating parts are assembled. There are loose fit (1), free fit (2), close fit (3), and very tight fit (5).

Standard Bolts

Bolts are classified according to the shape of their heads and nuts, and according to the quality of the material they are made from. Manufacturers produce the following kinds of bolts based on the American standards:

1. Regular square
2. Regular hexagon
3. Heavy hexagon
4. Regular semi-finished hexagon
5. Heavy semi-finished hexagon
6. Finished hexagon
7. Heavy finished hexagon

These bolts are represented on a drawing by a conventional way accompanied by a specification note. For example: 7/8 - 19 UNF - 2A x 3 BRASS FIN. HEX. HD. The first part (7/8 - 19 UNF - 2A) specifies the thread and the second specifies the bolt itself. 3, in this example, is the length of the bolt in inches, BRASS is the material from which it is made (if nothing is mentioned about the material then it is steel), and FIN. HEX. HD is the kind of bolt; in this case, it is finished hexagonal head bolt. The nut has the same shape and finish as that of the head but slightly thicker.

Screws

There are many kinds of screws in the market but the most important ones are: Cap screws, Machine screws, Set screws and Wood screws. Some of these are not represented at all on a drawing and some are represented by conventional symbols.

Rivets

These are cylindrical pins that have heads on one end only. Their purpose is to make permanent joints between plates or rolled sections. A head on the other end of the rivet is formed by hammering after it is driven into the members it is joining. Rivets might have different heads but they all function the same.

PROBLEMS

Solve the problems assigned by your instructor on a separate sheet of paper.

1 - Draw a Unified National (V-type), external, right hand, single thread and its cylindrical mating part (internal thread), having the following specifications:

External Thread

Nominal Diameter	= 50 mm.
Pitch	= 6 mm.
Threaded Length	= 90 mm.
Total Length	= 120 mm.

Internal Thread

Thickness	= 40 mm.
External Diameter	= 70 mm.

2 - Draw an external, right hand, single, square thread having the following dimensions:

Diameter	= 60 mm.
Pitch	= 20 mm.
Threaded Length	= 100 mm.
Total Length	= 140 mm.

3 - Repeat No. 2 for an ACME thread.

4 - Draw a regular hexagon head bolt and nut having the following specifications:

Diameter	= D = 20 mm.
Pitch	= P = 4 mm.
Threaded Length	= L = 60 mm.
Length	= l = 100 mm.

CHAPTER 10

DIMENSION LIMITS, TOLERANCE AND ALLOWANCE

Machines are usually made up of several parts which must fit together well otherwise they will not operate properly. The drawings that show the relationship between the different parts of a certain machine are called "Assembly Drawings".

The designer should know this relationship and dimension the parts accordingly. Since no part could be manufactured economically to exact dimensions, the designer should specify the variation in the dimension of a certain part without affecting its operation. Thus, one should put down its maximum and minimum dimensions which are referred to as "DIMENSION LIMITS". The difference between these two limits is called "TOLERANCE". This variation in dimension will give the workman at the shop some leeway in manufacturing a certain part. The smaller this variation is, the more expensive is the part. That is, tolerance, in a way, determines the cost of a part.

In order for two cylindrical mating parts in an assembly to fit well within each other, there should be some clearance between them depending on the degree of fit required. So, this clearance which is the difference in the dimensions of two mating parts is called "ALLOWANCE". Since each part has two dimension limits (maximum and minimum), one should calculate the allowance when the worst case prevails. This will occur when the hole size is at its minimum and the shaft size is at its maximum. A hole and a shaft are considered good examples of two cylindrical mating parts. Therefore:

Allowance = Minimum dimension of hole - Maximum dimension of shaft

Two methods are used in calculating the limits, the tolerances and the allowance of two cylindrical mating parts. These are the **Basic Hole** and the **Basic Shaft** methods. In the former method, the calculated size of the hole is considered as the basic one and the size of the shaft is obtained by subtracting the allowance from that size. In the other method, the size of the shaft is considered as the basic one and that of the hole is calculated by adding the allowance. Since standard tools are used in making the holes while shafts are done usually by grinding, it is preferable to use the **basic hole** method. Standard limits used in this connection are supplied by fit tables which could be found in graphics books. These tables are designed to be used when the **basic hole** method is employed. In order to calculate the limits by the **basic shaft** method, the same tables could be used along with the following formula:

Limits by Basic Shaft Method = Limits by Basic Hole Method ± Allowance

Allowance is positive in case of clearance and negative in case of Interference. The values of the tolerances and the allowance are the same in both methods. Note that tolerance is always positive while allowance could be positive, if minimum size of hole is greater than maximum size of shaft (clearance case), and negative, if minimum size of hole is smaller than maximum size of shaft (interference case).

Problem

Complete the following table by calculating whatever is missing using the **tables of fits**. These problems show the dimensions of two cylindrical mating parts.

PROBLEM NUMBER	1	2	3	4	5	6
BASIC SIZE (INCHES)	1.6422	1.9685	23197		3.1500	
CLASS OF FIT	RC3	LN2	LC4	FN4	LT1	RC8
METHOD OF CALCULATION	BASIC HOLE	BASIC SHAFT	BASIC HOLE	BASIC HOLE	BASIC SHAFT	BASIC SHAFT
LIMITS OF HOLE SIZE				2.7530 2.7518		
LIMITS OF SHAFT SIZE						0.7564 0.7544
TOLERANCE ON HOLE						
TOLERANCE ON SHAFT						
ALLOWANCE						0.0045

CHAPTER 11

DESCRIPTIVE GEOMETRY

Descriptive geometry is the study by which one makes use of the space relationships that exist between the different elements which make up solids to solve engineering problems graphically. These solutions take much less time than the analytical ones and produce results with adequate accuracy. This study requires a fair knowledge of space or solid geometry since it depends, to a great extent, on visualizing three dimensional objects in space. It makes extensive use of auxiliary views which were studied earlier. That is, finding the true length of a line or its end view and the true shape of a plane surface or its edgewise view, are essential operations in descriptive geometry. Some of the problems that may be solved graphically in this respect are the following:

1. **Slope of a Line**

The slope of a line is the acute angle that this line makes with the horizontal plane. It is usually designated by the Greek letter θ. It is positive when the line slopes upward and negative when it slopes downward. In some cases, slope could be given in percent as is the case in roads and canals, or It could be given as a ratio of horizontal distance to vertical one for example the sides of drainage ditches and embankments of highways and railroads.

2. **Slope of a Plane**

The slope of a plane is the acute angle that the plane makes with the horizontal. It also has a sign which is positive if it slopes upward and negative if it slopes downward. In geology problems, they refer to the slope as the "dip of a plane" since it is always downward.

3. **Bearing of a Line**

The bearing of a line is the acute angle between the horizontal projection of that line and the reference meridian (north-south line). It may be measured from either the north or the south towards the east or the west. Bearings represent one system of defining the direction of a line. Another system is the azimuth of a line.

4. **Azimuth of a Line**

The azimuth of a line is the angle that the horizontal projection of that line makes with the reference meridian measured clockwise from either the north or the south. Engineers and surveyors generally use the north as their reference point while astronomers use the south.

5. Strike of a Plane

The strike of a plane is the bearing of a horizontal line in the plane.

6. Parallel Lines

Two lines are said to be parallel when they lie in the same plane and do not meet at all (or meet at infinity). In engineering drawings, lines are parallel when their respective projections are parallel.

7. Intersecting Lines

Two lines are said to be intersecting if they have one point in common. In projection drawings, the front and side views of the point of intersection as well as the front and top views must be in horizontal and vertical alignment with each other respectively.

8. Perpendicular Lines

If the angle between two intersecting lines is 90°, then these lines are said to be perpendicular to each other. The true size of this angle will show in the view where one of the lines appears in its true length whether that view is a principal or an auxiliary one. Also, two lines are perpendicular (orthogonal) to each other if they do not intersect but one of them lies in a plane perpendicular to the other.

Finding the resultant of concurrent coplanar forces acting on a certain body is another problem which could be solved graphically (Vector Diagrams).

② Which is longer: XZ or YZ and by how much?

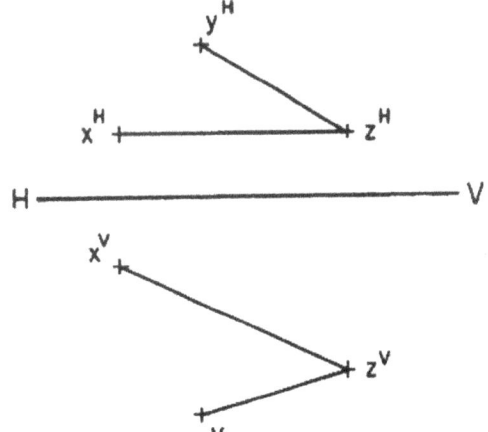

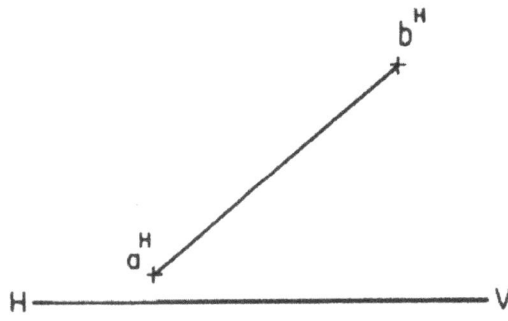

① Find and label the following for line AB:
- True Length =   mm.
- Bearing   =
- Slope   =

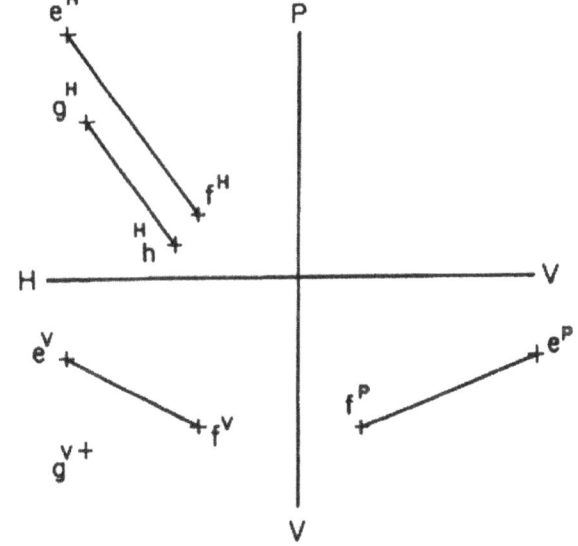

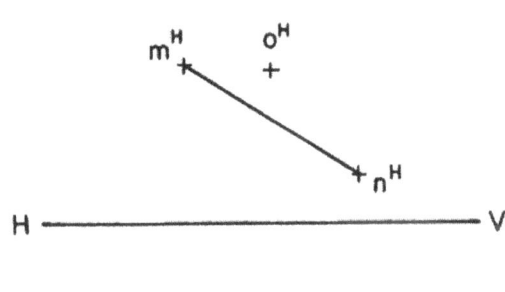

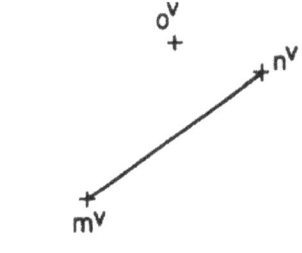

③ Construct line OP 20 mm. long and parallel to MN. OP lies in the south east quadrant.

④ EF and GH are two parallel lines. Draw the front and side views of GH.

TITLE:	
DRAWN BY:	
CHECKED BY:	
SCALE:	DATE:
SECTION:	DRAWING No.:

① Draw the profile view of CD which intersects AB at right angles.

② Find the shortest distance between the two horizontal lines PG and RS.

Dist =   mm.

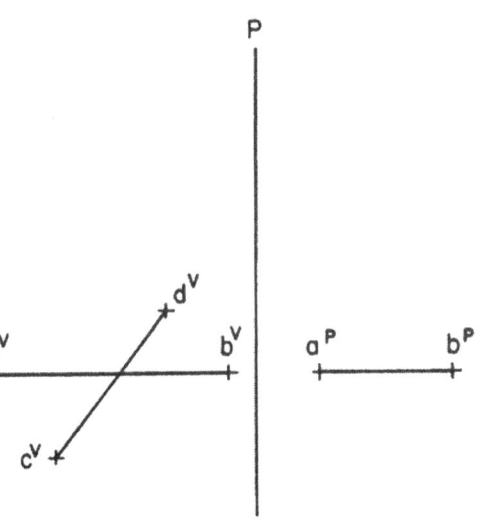
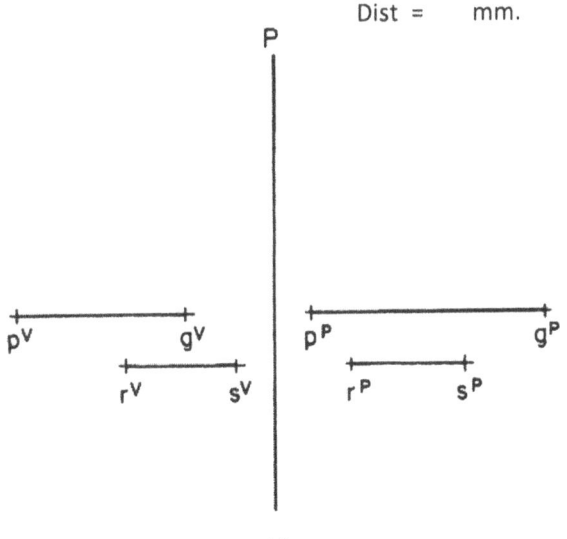

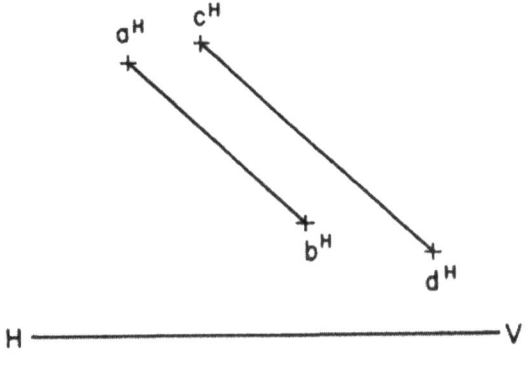
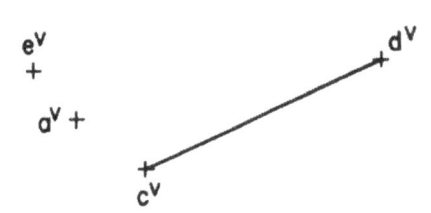

④ Construct line OP to intersect line MN at P, 20 mm from M. Also, find: T.L. of OP =   mm.
Brg. OP =   mm.

③ AB and CD are two parallel lines. Draw a horizontal line EF, 30 mm. long intersecting AB at M and having a bearing of N45° E.

TITLE:	
DRAWN BY:	
CHECKED BY:	
SCALE:	DATE:
SECTION:	DRAWING No.:

① Find the perpendicular distance from P to AB.   Perp. dist. =   mm.

② Find and label the true length of EF, its bearing and its slope.
T.L. =   mm; Brg. =   mm; slope =
Find the same for FE:
T.L. =   mm; Brg. =   mm; slope =

③ Find the Perpendicular distance from X to GH.
Dist. =   mm.
Find the slope and bearing of GH.
Slope =   ; bearing =

④ Find perpendicular distance from X to MN.
Perpendicular dist. =   mm.

Also, find:
Bearing MN =
Azimuth MN =
Slope MN =

TITLE:	
DRAWN BY:	
CHECKED BY:	
SCALE:	DATE:
SECTION:	DRAWING No.:

① Construct line AB having been given the following:
Length = 300 m
Bearing = S75°W
Slope = 30° downward
Use scale 1 cm = 50m.

$+ a^H$

H ——————————————— V

$+ a^V$

② A man climbing a mountain starts at point X and moves due north on a 10% slope path. After he reached point Y, 500 m. away, he turned right and started moving on a horizontal path having a direction of N25°E until he reached point Z, 300 m from Y where he took a rest.

Draw the projections of the path the man followed from X to Z.

Use scale 1:10,000

$x^H +$

H ——————————————— V

$x^V +$

TITLE:	
DRAWN BY:	
CHECKED BY:	
SCALE:	DATE:
SECTION:	DRAWING No.:

① Point C is connected to Points A and B by two straight ramps as shown in the diagram. Two cyclists Participating in a rally are heading towards C. One is going to use ramp AC and the other BC.
If they hit points A and B at the same time and continue towards C at an equal velocity, find:
- who is going to reach point C first and why?
- slopes of ramps AC and BC.
Label all your answers.

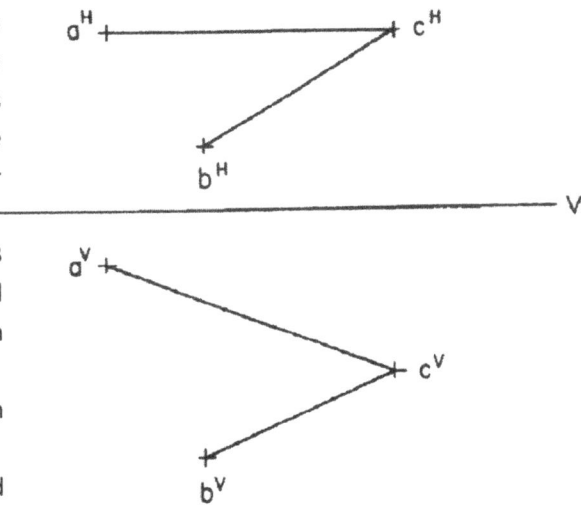

② An airplane in the process of taking off, starts moving from point A on the runway at maximum acceleration until it reaches point B, 3 km away where it takes of at an angle of 20° with the horizontal. The runway is horizontal and has a bearing of N60°E at point A. The airplane continues to have the same bearing as that of runway after taking off for a distance of 2.5 km (call this point C). It then makes a turn. Draw and label the trajectory followed by the plane from starting point A to point C, passing by point B using scale 1 cm = 1 km.

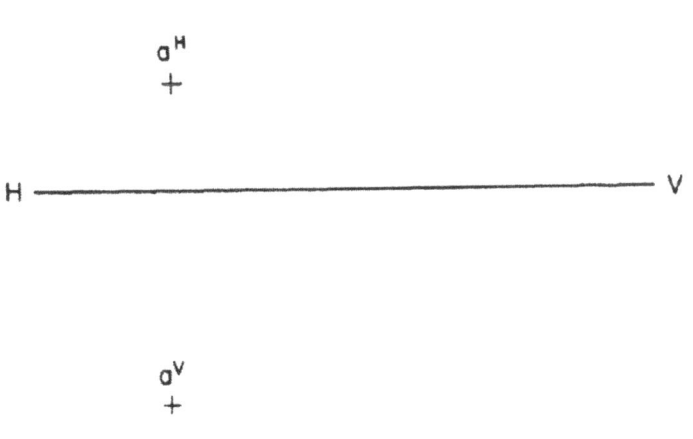

TITLE:	
DRAWN BY:	
CHECKED BY:	
SCALE:	DATE:
SECTION:	DRAWING No.:

1. Draw, on a separate sheet, a regular hexagon ABCDEF with AB having a length of 70 mm and a bearing of N 70° W. Measure and tabulate the bearings of all of the sides. Use scale 1:1.

2. The sides of a piece of land form a clockwise polygon ABCDEFGHA which is referred to in the surveying language as a traverse. A surveyor measured the sides and the interior angles of this polygon and found them to be:

AB = 44.9 m	Angle A = 109° 50'
BC = 38.1 m	Angle B = 70° 10'
CD = 49.3 m	Angle C = 216° 50'
DE = 40.0 m	Angle D = 123° 35'
EF = 26.5 m	Angle E = 28° 45'
FG = 28.2 m	Angle F = 297° 40'
GH = 48.8 m	Angle G = 92° 27'
HA = ?	Angle H = ?

If the bearing of AB is N 17° 00' E, draw, on a separate sheet, the above polygon to a scale of 1:1000 and measure the bearing and the azimuth from the north of every line. Tabulate the results. Find out the length of HA and the value of the angle H which the surveyor forgot to measure.

What is the relationship between lines BC and AH?

① Draw and label the true slope of ΔABC.

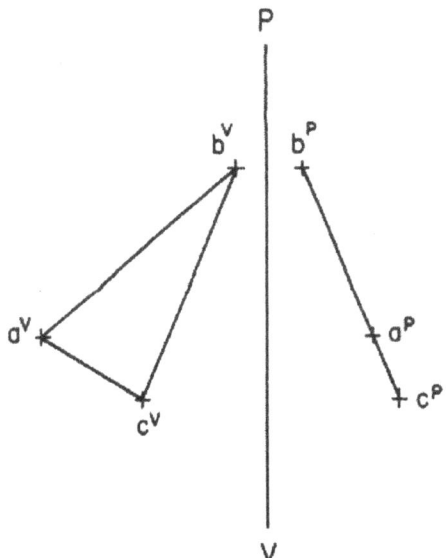

② Draw and label the front view of octagon ABCDEFGH.

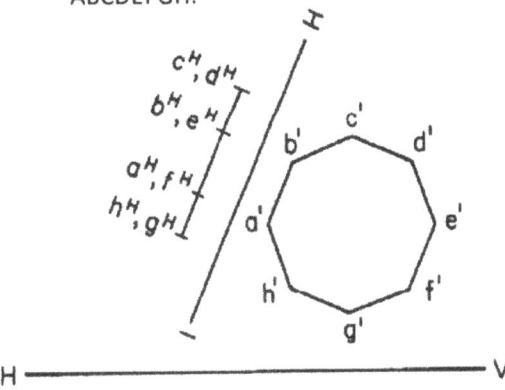

③ Draw and label the true shape of ΔXYZ.

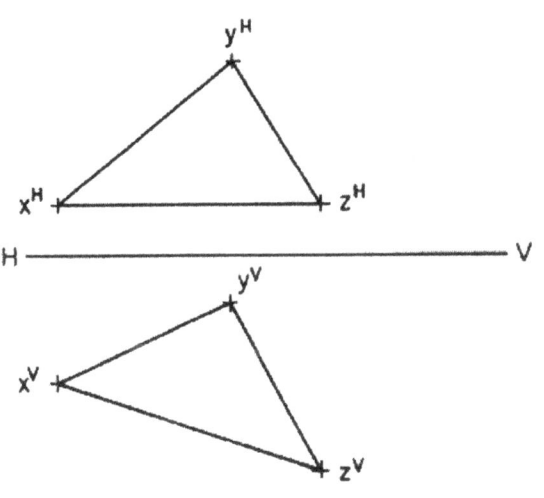

④ Find and label:
Strike of plane EFG =
Slope of plane EFG =

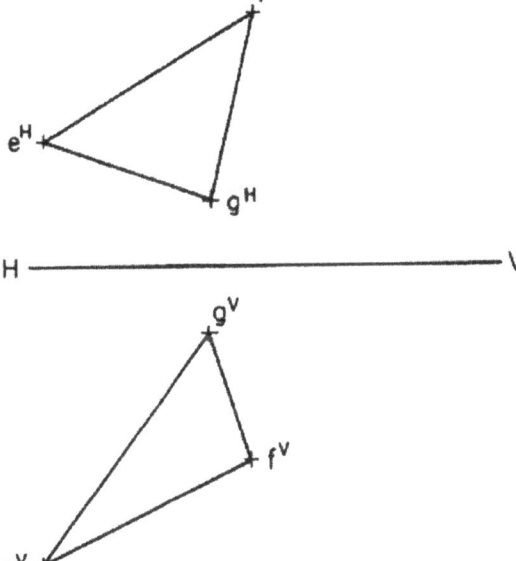

TITLE:		
DRAWN BY:		
CHECKED BY:		
SCALE:		DATE:
SECTION:		DRAWING No.:

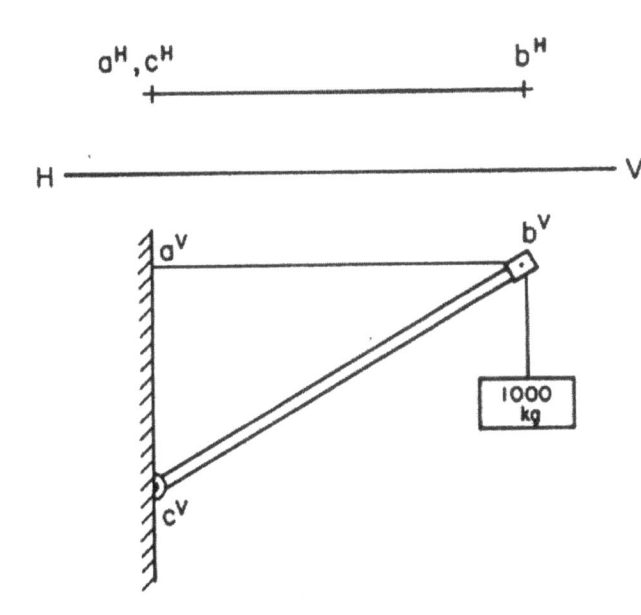

① Find the forces in the cable AB and the boom BC. Give the results in Newtons (a mass of 1 Kg results in a force of 9.807 Newtons). Indicate tension by (+) and compression by (-). Scale of vector diagram:
10 mm. = 3000 Newtons.

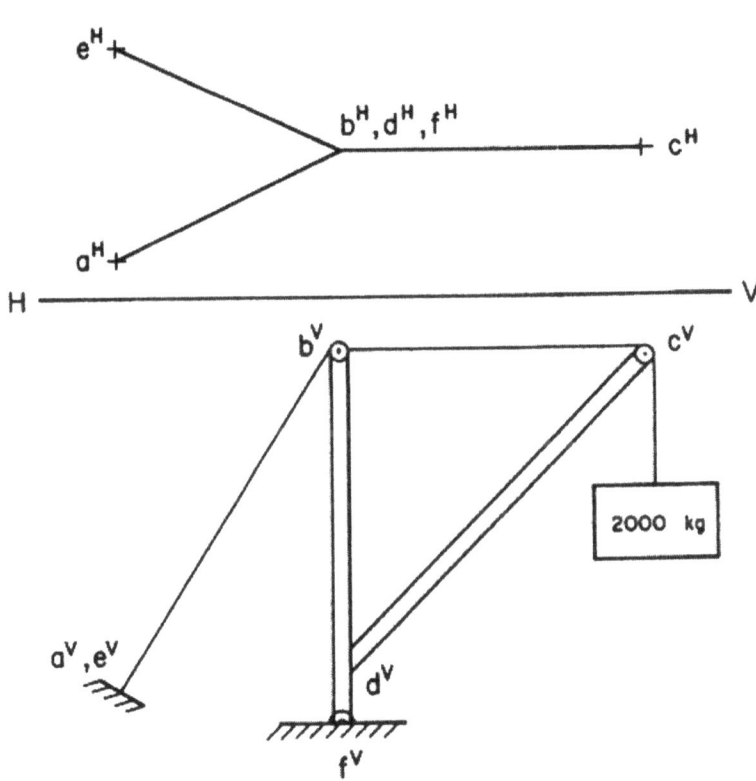

② Find the forces in boom DC cable BC, mast BF and the guy wires BA and BE.
Scale 10 mm. = 5000 Newtons

TITLE:	
DRAWN BY:	
CHECKED BY:	
SCALE:	DATE:
SECTION:	DRAWING No.:

CHAPTER 12

INTERSECTIONS AND DEVELOPMENT OF SURFACES

I. Intersections

The problem here is to find and define the line joining all of the points common to two surfaces known as the line of intersection. Generally, the method used in this connection consists of passing cutting planes through both surfaces and finding the point of intersection of the elements cut off. The process is repeated as many times as needed to establish a number of points which if joined in the right order with the correct visibility form the required line of intersection. In order to find the true length or shape of the line of intersection, auxiliary views are employed.

II. Development of Surfaces

Solids, generally, are bounded by either plane surfaces such as cubes and pyramids or by single curved surfaces such as cylinders and cones, or by double curved surfaces such as spheres and paraboloids, or by warped surfaces such as helicoids and conoids. Laying out flat patterns from which these surfaces could be made without stretching the material is called "Development". It is worth noticing that each element of the development should be equal to the true length of the corresponding element on the real figure.

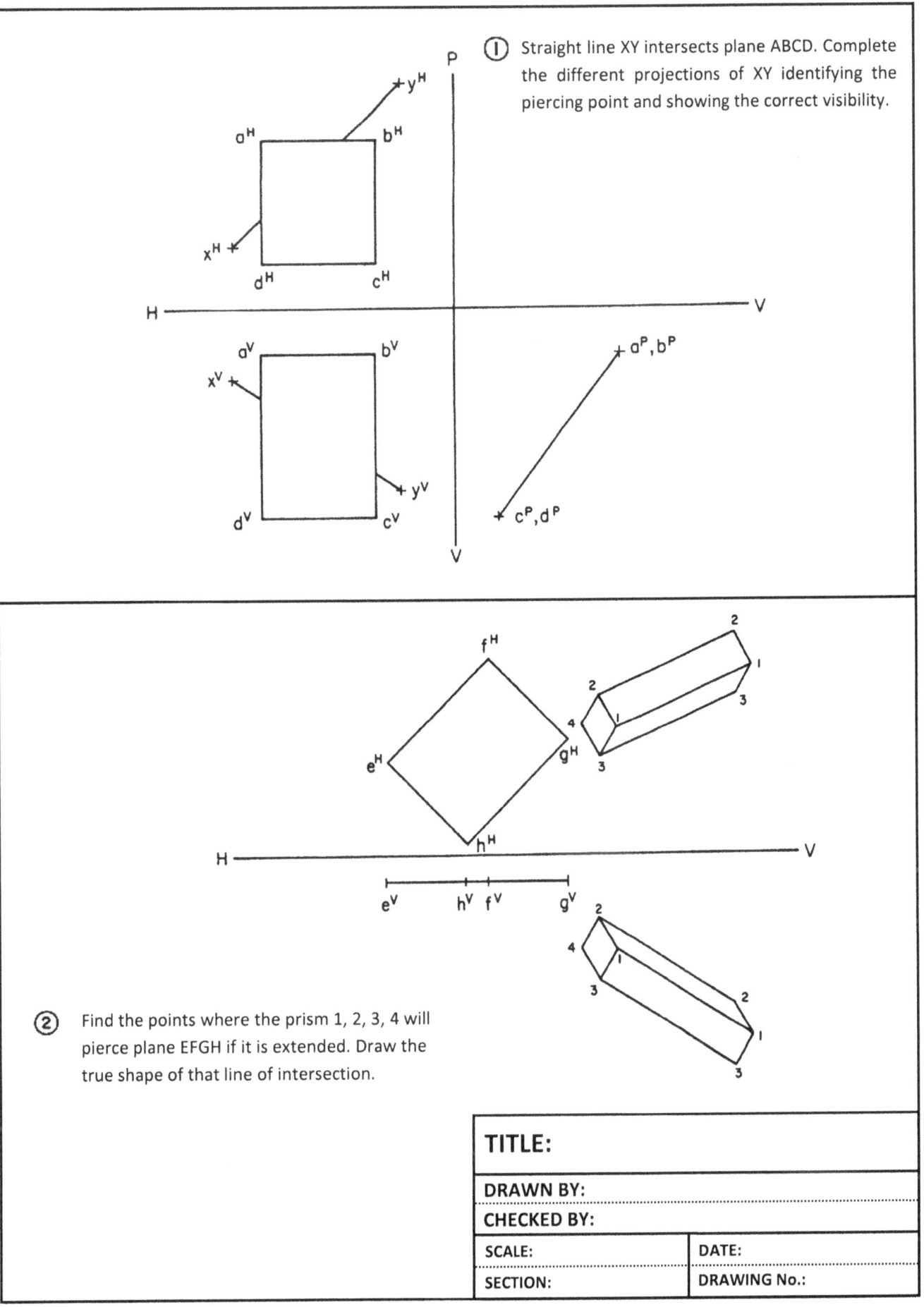

① Draw the line of intersection of the two planes ABC and EFGH showing the correct visibility.

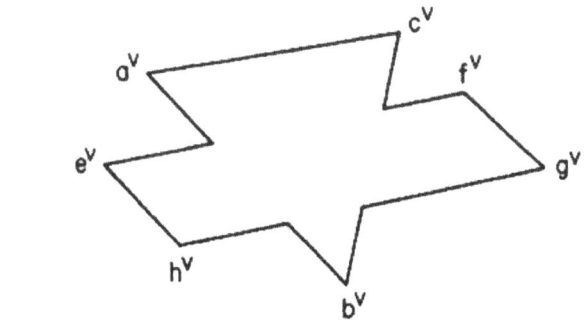

② Find the exact shape of the opening to be made in the roof of the house to allow the installation of the circular chimney shown above.

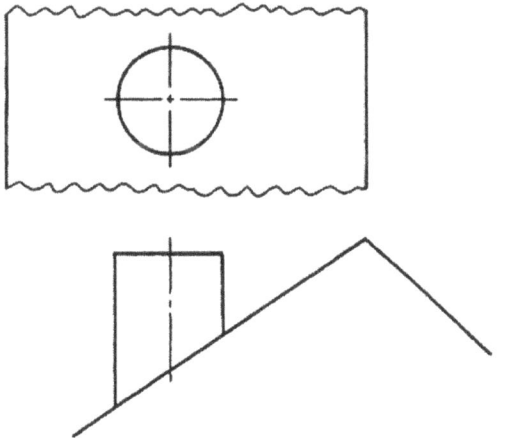

TITLE:	
DRAWN BY:	
CHECKED BY:	
SCALE:	DATE:
SECTION:	DRAWING No.:

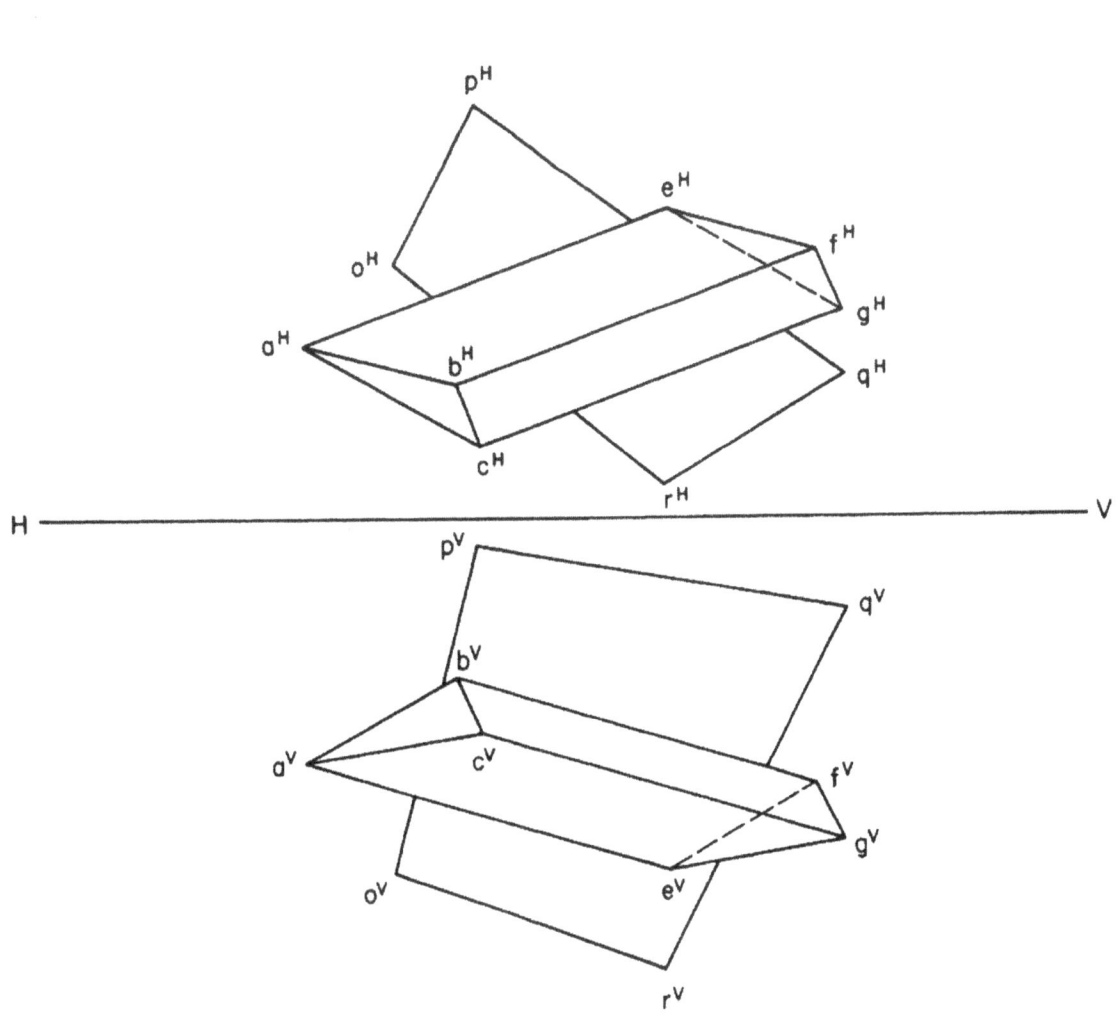

- Draw the line of intersection of the triangular prism ABCEFG and the plane OPQR showing the correct visibility
- Draw the true shape of that line of intersection

PROBLEM

SPIRAL STAIRCASE

Draw the plan and elevation of a spiral staircase connecting two floors in a house. The stairs are cantilevered from a central circular concrete column and arranged in such a manner that they make a complete circle by going around the column from one floor to the other. The following specifications are given:

Height between the two floors	= 320 cm.
Column diameter	= 100 cm.
Stair length	= 120 cm.
Number of stairs required	= 16

Neglect the thickness or height of each stair, just draw the profile of the staircase which has the form of a right helicoid. Show the proper visibility by imagining that the column is to the left hand side of the person climbing up from one floor to another. Use scale 1:40.

N.B: Do not forget to draw the line of intersection between the stairs and the column.

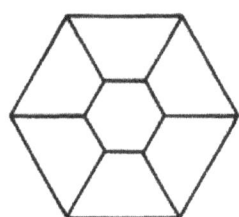

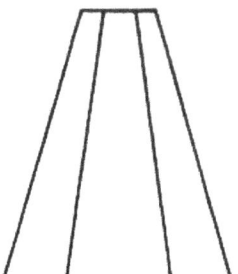

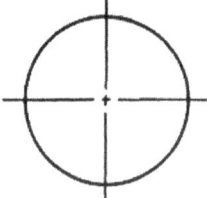

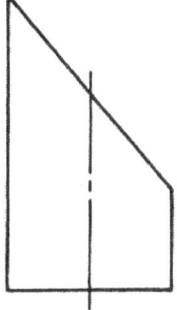

Draw the development of each of the two surfaces shown above. 1 is a frustum of a hexagonal pyramid and 2 is a frustum of a right cylinder.

TITLE:	
DRAWN BY:	
CHECKED BY:	
SCALE:	DATE:
SECTION:	DRAWING No.:

①

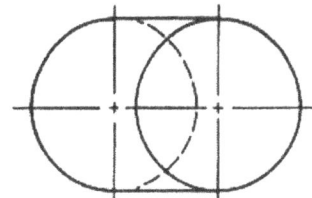

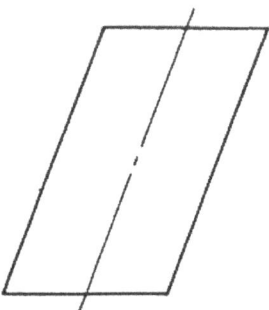

 ②

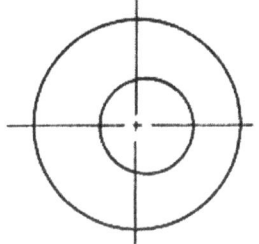

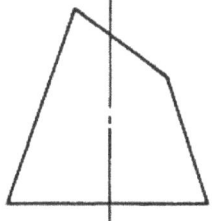

Draw the development of the inclined cylinder shown in 1 and the frustum of the cone shown in 2.

TITLE:	
DRAWN BY:	
CHECKED BY:	
SCALE:	DATE:
SECTION:	DRAWING No.:

CHAPTER 13

INTRODUCTION TO AUTOCAD

AutoCAD is one of the favourite software programs available in the market that is commonly used in Computer Aided Design (CAD). Typically CAD software is required to produce and present the results of building drawings, computer models, structural or mechanical parts, or finalized construction details and shop-drawings. Nowadays, a draftsman should possess the knowledge to work efficiently with this type of software to produce drawings according to the required drawing standards.

AutoCAD software, which is the focus of this chapter, is evolving every year to accommodate the needs of the users. The software producers also improve the performance of the program to take better advantage of computing capabilities as computers get faster every day. However, this does not change the fact that the basic techniques of the program remain the same and that knowledge of these basic techniques can be acquired. The next step for the user would be to adapt to the latest version used in his/her workplace to utilize AutoCAD's abilities to the fullest.

AutoCAD Interface

AutoCAD is loaded with certain default toolbars and features which open when the software is run for the first time. These toolbars usually cover three sides (right, left and top) of the screen and the bottom part is normally left for certain buttons and the program dialogue line. This line is used to communicate with the user and it also shows the available options for the selected command. It should be noted that all the toolbars can be customised based on the user's needs and several additional toolbars could be loaded at any time. A sample interface of the program is illustrated in the figure below.

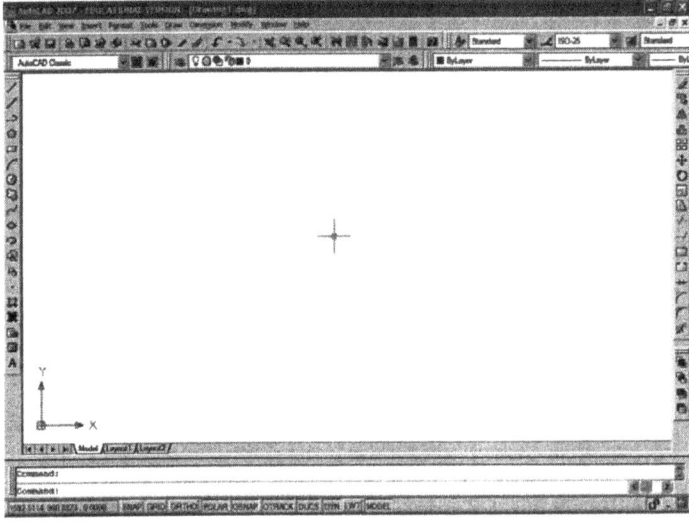

Some of the basic toolbars such as standard and layers located at the top of the screen are used to perform basic file management tasks like opening or saving a project. An AutoCAD file is saved with a **.dwg** extension and could be used in many other compatible programs. Other toolbars, located at the sides of the screen, are used to create or modify drawings. All the commands, including the ones already available in the toolbars and their special features, can be accessed directly through menus.

Before starting a drawing, certain preparations are needed. Some of the most common requirements are explained in the following section.

Screen Background Colour

In a 2D AutoCAD drawing, which is commonly used by engineers and drafters, the background colour is normally set to either black or white by default. The background colour and cursor size can be changed by going to the tools/options menu and then selecting the display tab. In that window the display setting of the program can be customized based on the user's preferences.

Drawing Units

Every drawing can be drawn using different units with variable precisions. In North America both the metric and the imperial systems of units are used commonly. The drawing will indicate the necessary precision to the craftsman in the machine-shop. For instance, if the length of a mechanical part is measured as 13.500 mm instead of 13.5 mm, then that indicates that the acceptable tolerance for the length is 0.001 instead of 0.1 (100 times more precise). In order to achieve this level of accuracy certain machines have to be used and in addition to that, extra labour hours are needed to finish the job. This will result in an exponentially more expensive part and a slower job. The proper units for the drawings can be set in the format/drawing unit menu in AutoCAD according to the designer's preferences.

Drawing Limits

The size of the drawing canvas can be set using the tools/drawing limits command. The program can extend the drawing beyond these set limits if needed and also the drawing limits can be adjusted at any time. It only affects certain options such as grid and snap.

Zoom and Pan

Zoom and pan commands are used commonly to navigate through different segments of the drawing. The zoom command, which is used to focus on a certain portion of the drawing, can be activated using the onscreen zoom toolbar or by typing **"zoom"** or just **"z"** in the command line. The same command can also be used to zoom out to have a better view of the entire

drawing. The following options are the most popular zoom options. The pan command is used to navigate through the screen.

Zoom\all:

This option is used to zoom out the drawing limits set by the user. If the drawing is bigger than the set limit, then AutoCAD will zoom out to display the entire drawing on the screen.

Zoom\extents:

This zoom option expands the objects to cover the entire display area. This will result in displaying all of the objects in their biggest possible size.

Zoom\free zoom:

The free zoom is simultaneously used with pan to zoom in and out of different segments of the drawing as well as to move to different areas of the screen.

Zoom\selected rectangular area:

This option is a popular command used usually after using the zoom all or zoom extents command to focus on a specific portion of the drawing selected within a rectangle by clicking the two diagonally opposite corners.

After completion of basic preparations, now is the time to focus on creating the main drawing. The **draw toolbar** is used to produce the basic shapes and the **modify toolbar** is needed to fine-tune the drawing to its finalized shape. In the following sections some of the common commands in these two toolbars are explained briefly.

Draw Toolbar

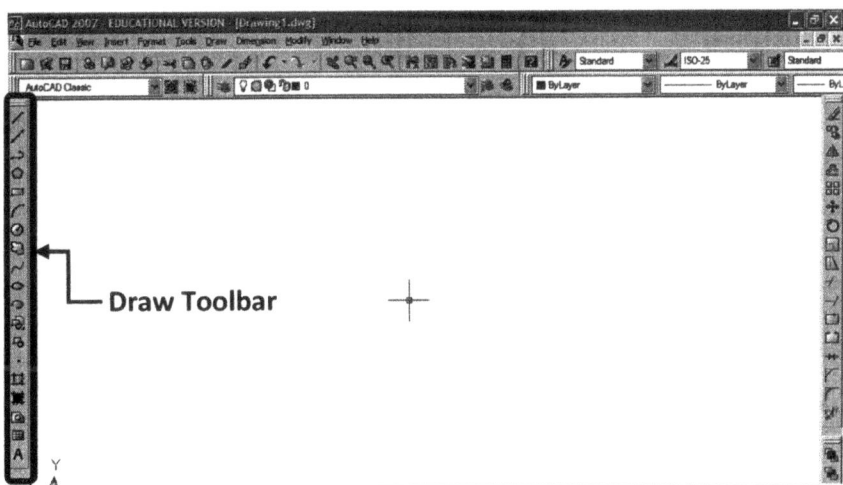

Line

The line command is used to draw a straight line from the starting point (i.e., point A) to an ending point (i.e., point B). After drawing the first line, the process could be finished by pressing either the Enter, Space Bar or Escape Key. If another set of coordinates is inserted instead or another point is clicked the line command will continue drawing straight lines. The coordinates of the start and end points can be given using the absolute coordinate system (meaning all the coordinates are given based on the origin (0,0) point) or relative coordinate system. In a relative coordinate system the user simply chooses the starting point of the line and then indicates the length and direction or coordinates of the second point based on the first point. AutoCAD software can distinguish the relative coordinates by the "@" sign inserted at the beginning of the coordinates. The difference between the two systems is expressed visually in the following figures:

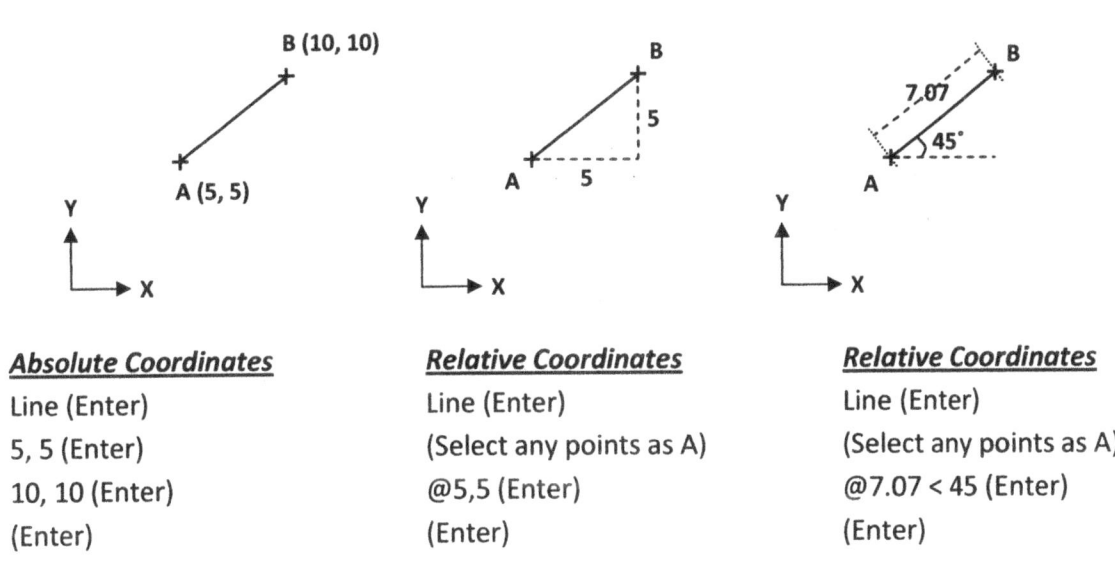

Absolute Coordinates	*Relative Coordinates*	*Relative Coordinates*
Line (Enter)	Line (Enter)	Line (Enter)
5, 5 (Enter)	(Select any points as A)	(Select any points as A)
10, 10 (Enter)	@5,5 (Enter)	@7.07 < 45 (Enter)
(Enter)	(Enter)	(Enter)

Note: in a line command "<" sign is used to indicate the angle of the line.

Construction Line

This command produces a line which continues indefinitely from both ends. These lines can be used as guidelines in the drawing to outline the rest of the drawing which is useful in producing orthographic and auxiliary views. One option is to erase these lines after using them. The other option is to move these lines to another layer and turn the layer off when it is not needed. The latter option could be useful since sometimes these lines are needed to do further

modifications to the drawings at a later time by turning that layer on again. The instructions on how to use layers are provided in the following chapter.

Arc

The arc command is used to create an arc of a circle. There are multiple ways to perform this command each suitable for a different situation. All different shapes of this command can be accessed through the draw\arc menu.

Circle

This command is used to draw circles. A circle can be drawn using center point and radius or diameter, or by using points on its perimeter.

Ellipse

The ellipse command is similar to the circle command except it is used to draw ellipses.

Hatch

This command is used to fill an enclosed area with a certain hatch pattern. The enclosed area can be defined by clicking inside an object with a closed perimeter (no discontinuity allowed) or by manually selecting the corners of the area.

The hatch pattern is selected from a series of pre-selected patterns provided by AutoCAD. Each pattern represents a certain material, finish or surface type. For example, a few of these hatch patterns and their meanings are presented in the figure below.

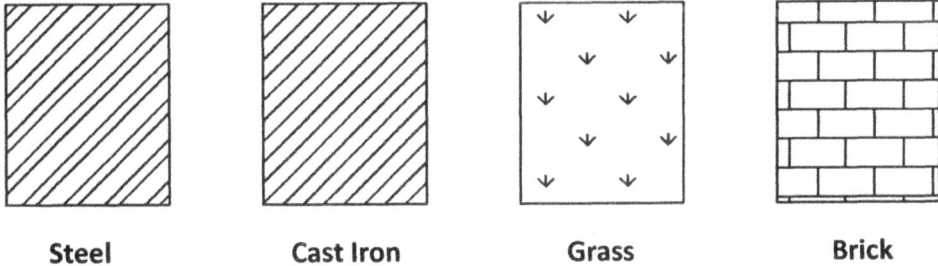

    Steel   **Cast Iron**   **Grass**   **Brick**

The scale and alignment of the pattern can be adjusted based on the size and specifications of the drawing.

Text

The text command is used to add comments and notes to the drawing to clarify the details.

Modify Toolbar

The modify toolbar is normally located at the right side of the screen as shown below. Some of the commands are explained briefly.

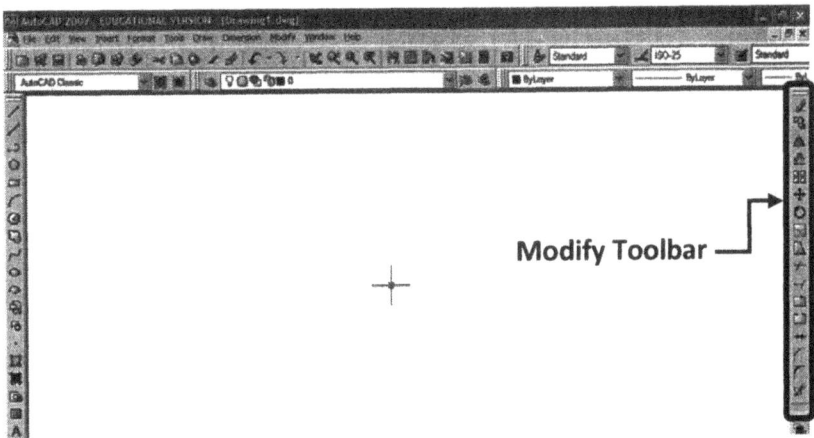

Erase

This command is used to remove unwanted objects (e.g., lines, circles, arcs, etc). Pressing the **Delete** button after selecting the objects produces the same result.

Copy

The copy command is used to duplicate an object at another location. This command gives the user the option to repeat copying the original object as many times as needed until the **Escape** or **Enter** Key is pressed.

Offset

This command is used to create a copy of the selected object at a specified distance. The offset command performs a certain task differently depending on the selected object. If the selected object is a line then the command will produce a parallel line with the exact same length at the selected side of the original line. In the case of working with an arc or circle, the offset command produces a similar arc or circle using the same center with a different radius. Some of these examples are shown below.

 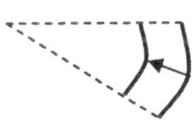 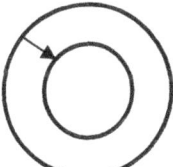 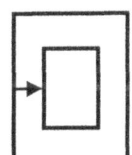

Array

This command is used to create multiple copies of the original object at once in a pattern. The newly created objects could be placed in a rectangular or polar array.

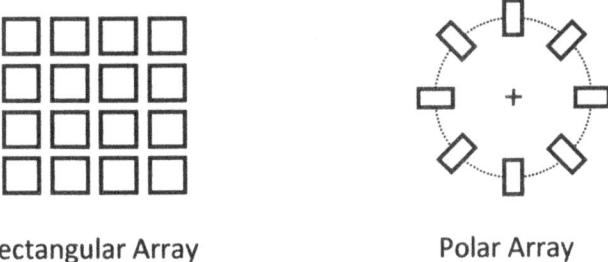

Rectangular Array   Polar Array

Move

The move command is used to relocate the object to another part of the drawing.

Rotate

This command is used to rotate an object at any given angle around a selected pivotal point. A positive angle will rotate the object counter-clockwise and a negative angle will turn the object clockwise.

Trim

The trim command is used to remove the extra portion of an object (e.g., line, circle, etc.) beyond a selected edge.

Extend

Extend is used to extend an object (e.g., line, arc, etc) up to a selected edge. Extend and trim commands are two powerful commands used together to modify the drawings.

Fillet

Fillet is used to produce rounds and fillets at the intersection of two lines. If the radius of the arc is selected as zero, then the lines are connected as straight lines. In other words, the fillet command will extend or trim the original lines to fit the rounded corner perfectly. Sample applications are shown in the following figure.

Radius "r"   Radius Zero

Layers

In a 2D drawing, sometimes an engineer has to show or hide certain details in a drawing to clarify certain features in different plots. A good example of this case can be seen in the floor plans of a house used by engineers. An architect may layout the outline of the structure in one layer and then a structural engineer will provide the location and the size of the structural elements in a different layer. Afterwards, a mechanical engineer will provide the details of the mechanical units (ducts, size and location of the heating and cooling systems) on a separate layer. An electrical engineer will use the same layout and will provide all the wiring detailing. Furthermore, if desired, the user can even furnish the floor area to have an idea how it will look when occupied.

After drawing the details in separate layers, the draftsman will decide what details to plot based on the needed details by turning off the unwanted layers. In other words, layers are very similar to transparent pages on top of each other where one has the option to remove or add a transparent if needed. In AutoCAD a toolbar is dedicated to layer management.

One of the advantages of the layers toolbar in AutoCAD is that the style or properties of all the elements within a layer can be modified by changing the layer properties in the management tool. For example, if the user wishes to use thicker lines or change the line style or colour, he/she can do that by simply changing that in the layer management tool instead of selecting and changing the properties of every individual element.

A new layer can be added through the layer management button located on the layers toolbar. Different layers can be turned off or on either through the management button or by using the quick drop-down menu located on the layers toolbar.

If the layer menu is not loaded into AutoCAD by default, the user can load the toolbar by right-clicking on any toolbar and selecting **Layers Toolbar** or by selecting tools/toolbars/layers in the toolbar menu. The layers toolbar is shown in the figure below.

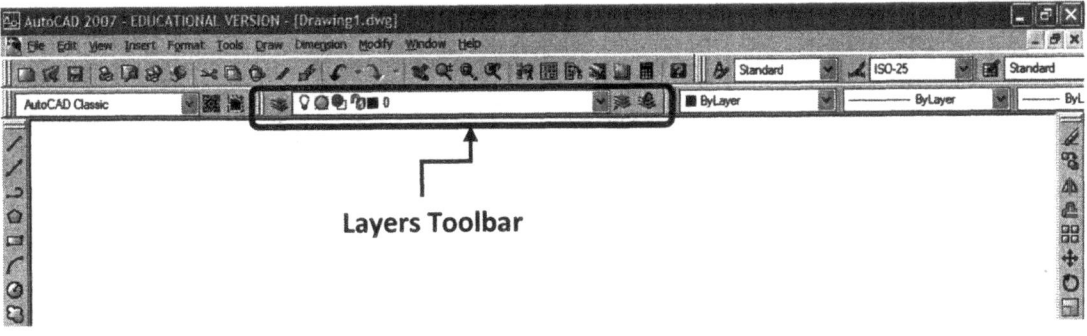

Layers Toolbar

Dimension Toolbar

This toolbar is not loaded by default and the user should load it by clicking on the toolbar menu and selecting **Dimension** or by selecting the tools/toolbar/AutoCAD/Dimension toolbar menu. After it has been loaded for the first time, AutoCAD will load this toolbar (or any other toolbar that has been added) automatically unless the user decides to close the toolbar again. Dimensions are normally placed in a separate layer in order to facilitate hiding/showing whenever needed and to expedite the adjustment process.

AutoCAD normally sets up a certain size for the letters and arrows and the dimensions to be displayed based on the initial size of the screen. However, in most cases, this size is not suitable for the drawing and needs to be modified. This can be done by accessing the Dimension/Dimension Style menu and then modifying the used dimension style. There are multiple options in different tabs of the dimension style menu, giving the user full control over the details on how the dimensions should be displayed. In the **Fit tab,** the **overall scale** option is provided to scale up or down the dimension size to match the size of your drawing. Another one of the favourite options in dimension style is located in the **Symbols and arrows tab** where the user can choose any **type of arrowhead** that is appropriate for the drawing. The rest of the options could be useful depending on the situation and should be explored by the trainee. The dimension toolbar is shown below.

Some of the most commonly used commands are explained briefly below.

Linear Dimension

This command is used to indicate the linear length of an object parallel to the X or Y axis. It is used often in providing the general dimension of any object.

Aligned Dimension

This command measures the length of an object (i.e., line) parallel to the object's main axis. This is useful when the overall length of an inclined object is needed.

Arc Length

The arc length command is used to measure the length of an arc section.

Radius

This command displays the radius dimension of a circle or an arc.

Diameter

The diameter command is used to measure the diameter of a circle or an arc.

Angular

The angular command is used to display the angle between two lines.

Baseline Dimension

This command is used to provide the dimension of different parts of an object relative to a selected reference line.

Continuous Dimension

The continuous dimension command is used to measure the distance between continuous points of an object.

Center Mark

Center mark is used to indicate the center point of an arc or a circle with a cross mark.

Page Layout

The final drawing should be plotted with proper scaling and sufficient information about it. This information usually includes the name of the project, the person or company who produced the drawing, the owner or sponsor of the project, drawing scale and any additional information which is deemed necessary. Before plotting, the drawing is framed within boundary lines and a small rectangular section at the lower right hand corner is used to provide the above information. In AutoCAD, the software has several pre-selected layouts to choose from. In addition, it gives the option of modifying or creating a personalized layout and saving it so it can be loaded for different projects.

The layout command can be loaded by clicking on one of the layout tabs at the bottom of the screen and selecting the option to produce a new layout or load an existing layout into the AutoCAD project file. The user could also access this command by typing "LAYOUT" directly in the dialogue box.

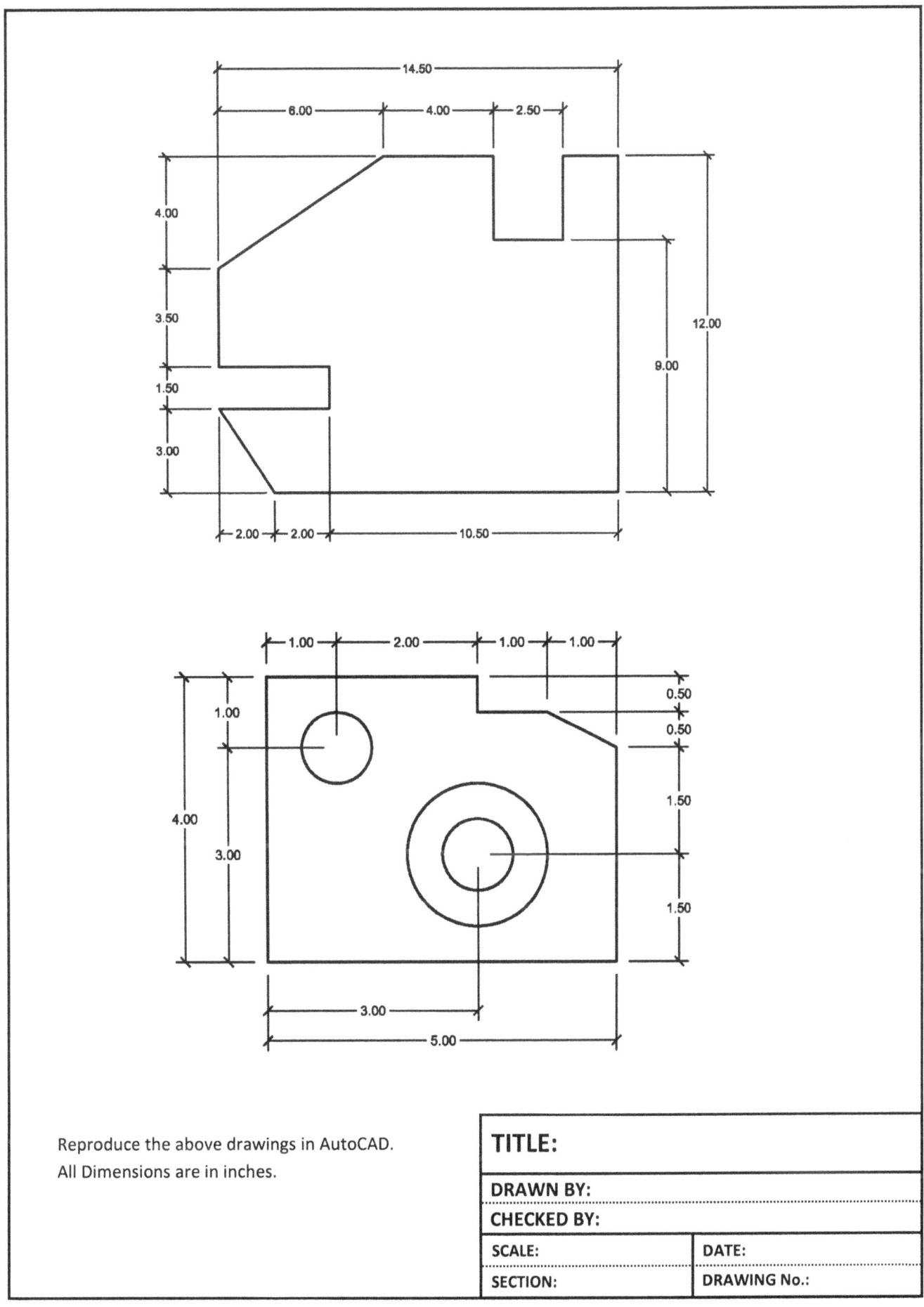

Reproduce the above drawings in AutoCAD.
All Dimensions are in inches.

TITLE:	
DRAWN BY:	
CHECKED BY:	
SCALE:	DATE:
SECTION:	DRAWING No.:

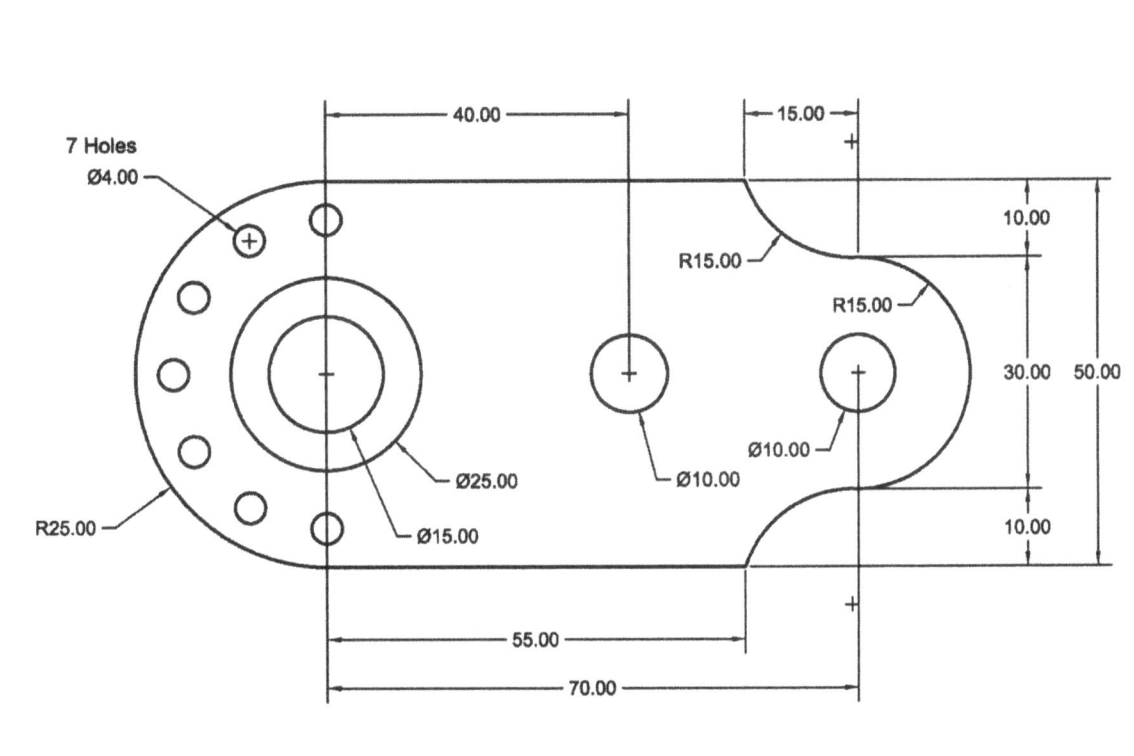

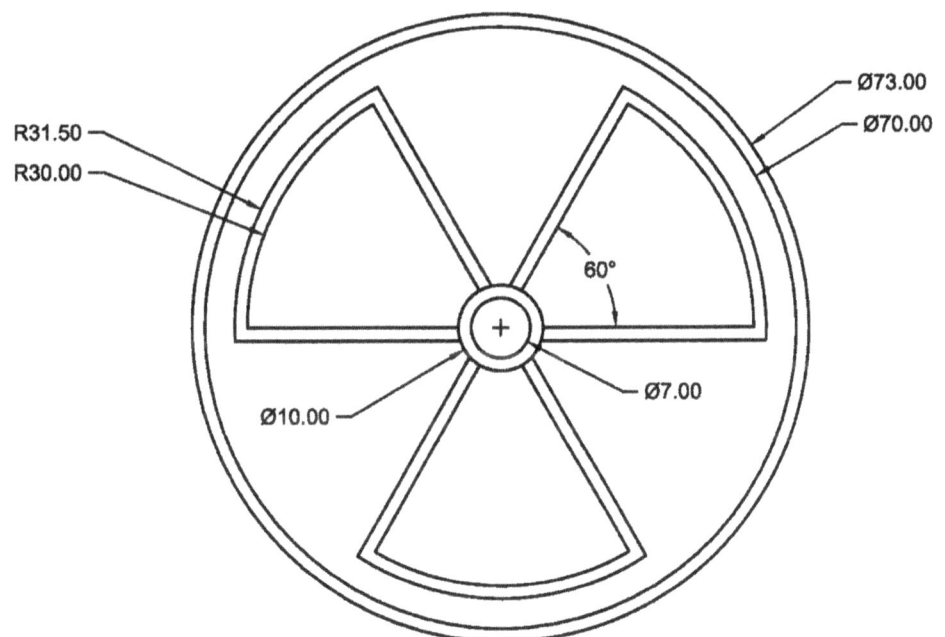

Reproduce the above drawings in AutoCAD.
All Dimensions are in mm.

TITLE:	
DRAWN BY:	
CHECKED BY:	
SCALE:	DATE:
SECTION:	DRAWING No.:

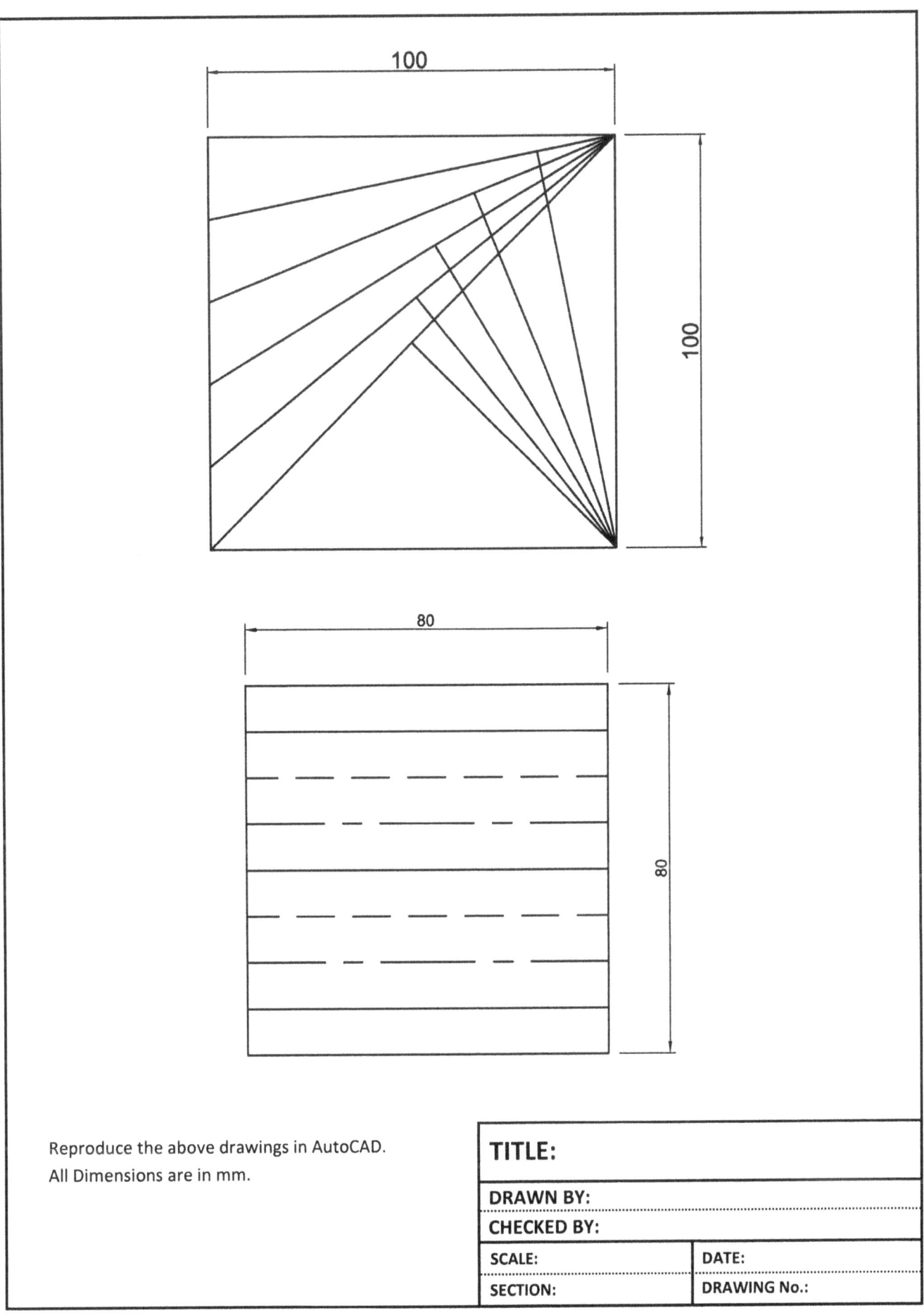

Reproduce the above drawings in AutoCAD.
All Dimensions are in mm.

TITLE:	
DRAWN BY:	
CHECKED BY:	
SCALE:	DATE:
SECTION:	DRAWING No.:

A Typical Wall Section

Usual Dimensions in a House:

- Interior wall thickness = 4" = 10 cm
- Exterior wall thickness = 6" to 12" = 15 cm to 30 cm
- Door height = 7 ft = 210 cm
- Door width = 3 ft = 90 cm
- Main entrance = 3.5 ft to 6 ft = 105 cm to 180 cm
- Window height above floor = 3 ft = 90 cm
- Height of the window = 4ft = 120 cm
- Width of the window = varies depending on the room size
- Staircase width = 3'- 4" = 100 cm
- Height of a stair = 6" to 7" = 15 cm to 18 cm.

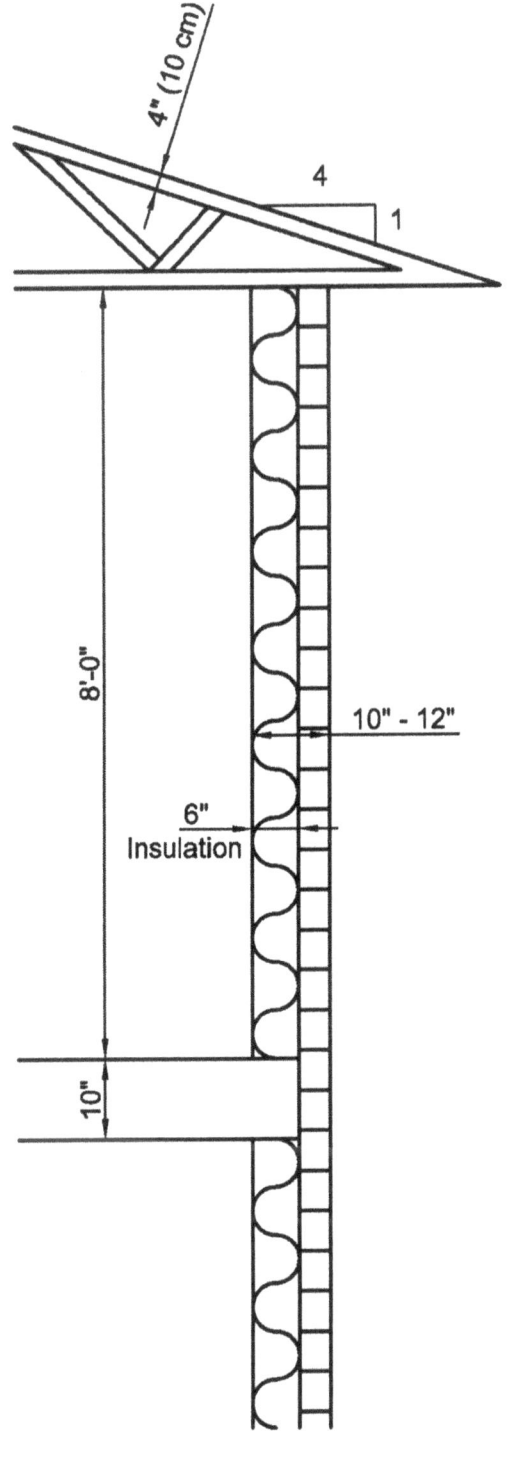

TITLE:	
DRAWN BY:	
CHECKED BY:	
SCALE:	DATE:
SECTION:	DRAWING No.:

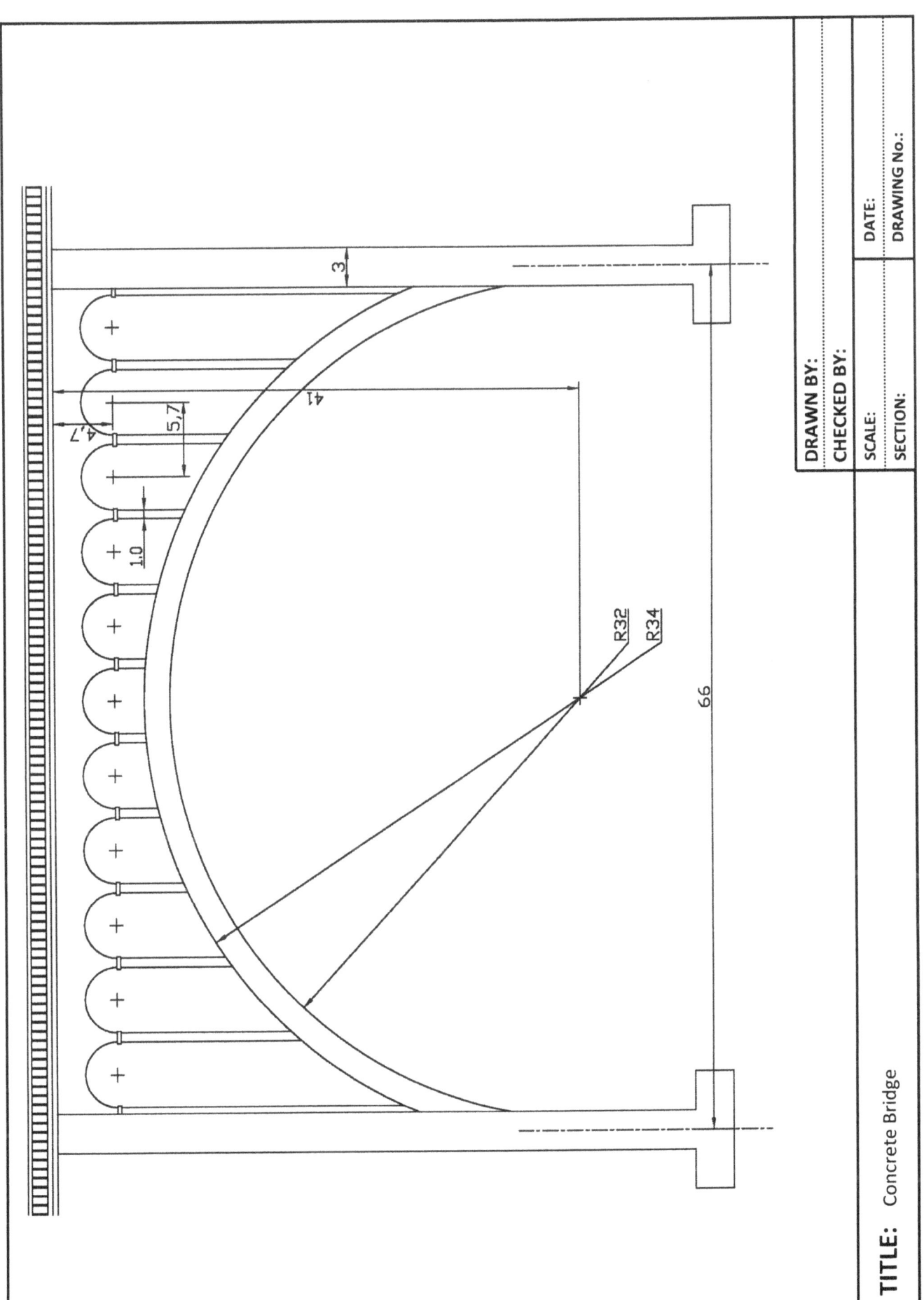

TITLE: Concrete Bridge

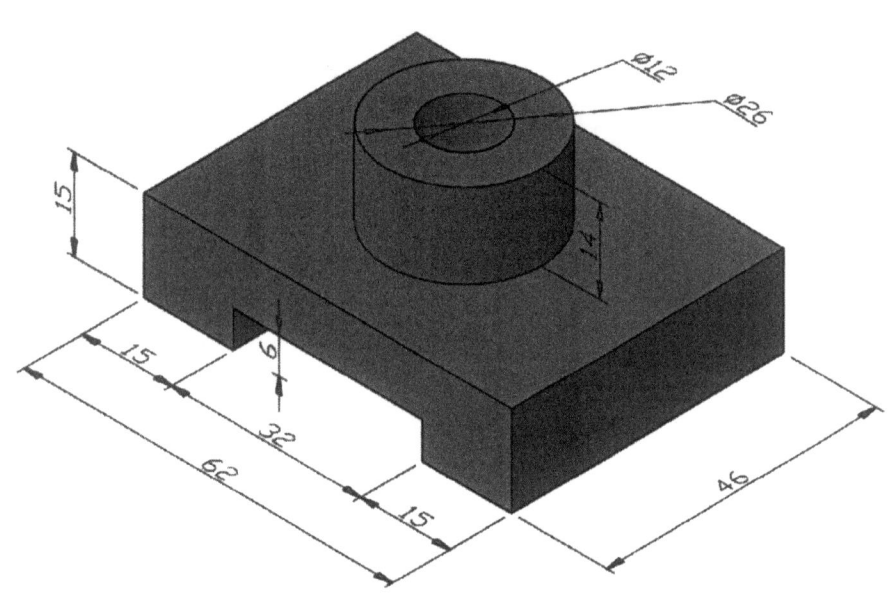

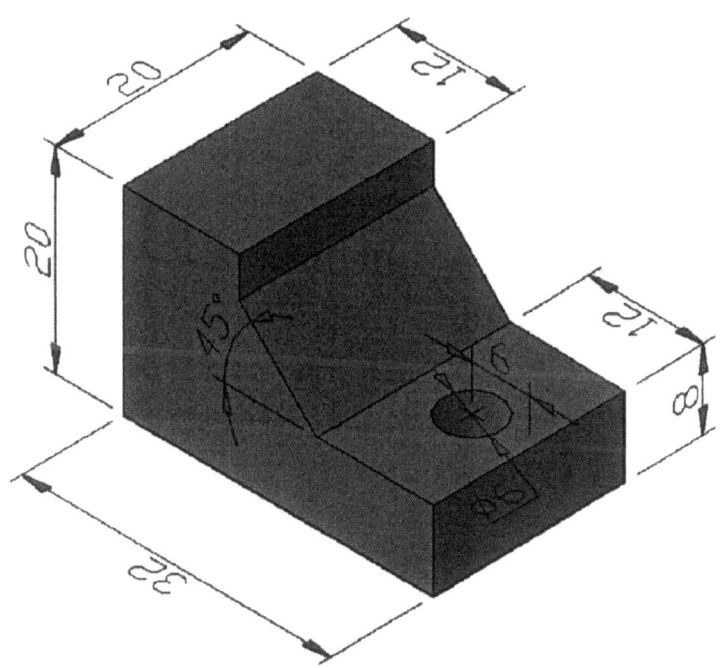

Draw the necessary views to represent the object using AutoCAD.
All the dimensions are in mm.

TITLE:	
DRAWN BY:	
CHECKED BY:	
SCALE:	DATE:
SECTION:	DRAWING No.:

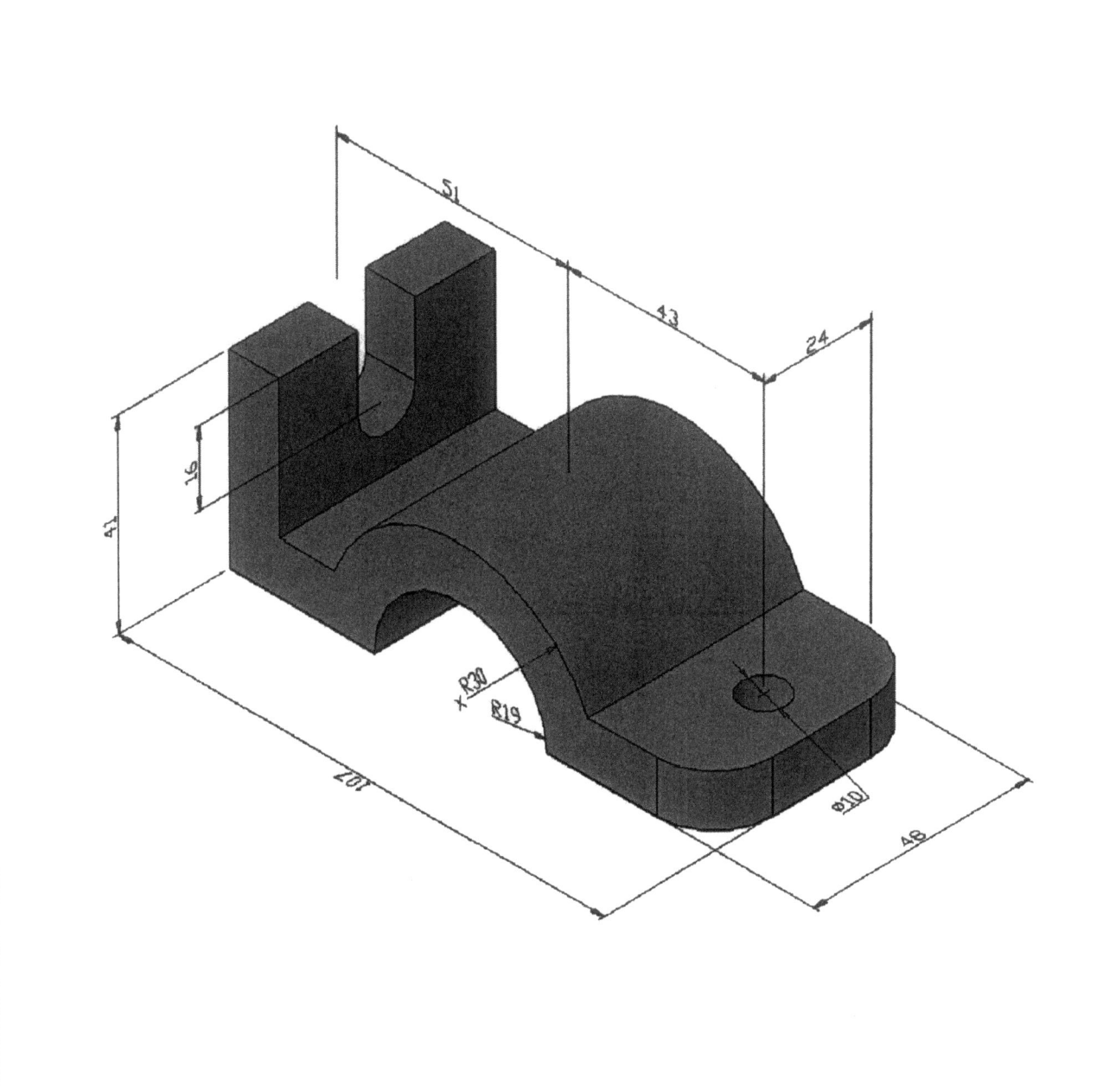

Draw the necessary views to represent the object using AutoCAD.
All the dimensions are in mm.

TITLE:	
DRAWN BY:	
CHECKED BY:	
SCALE:	DATE:
SECTION:	DRAWING No.:

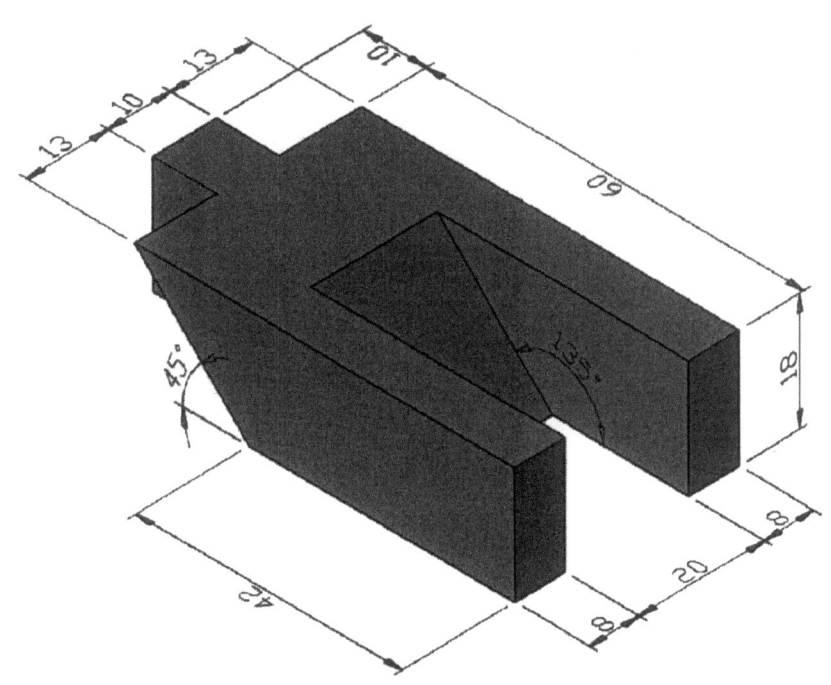

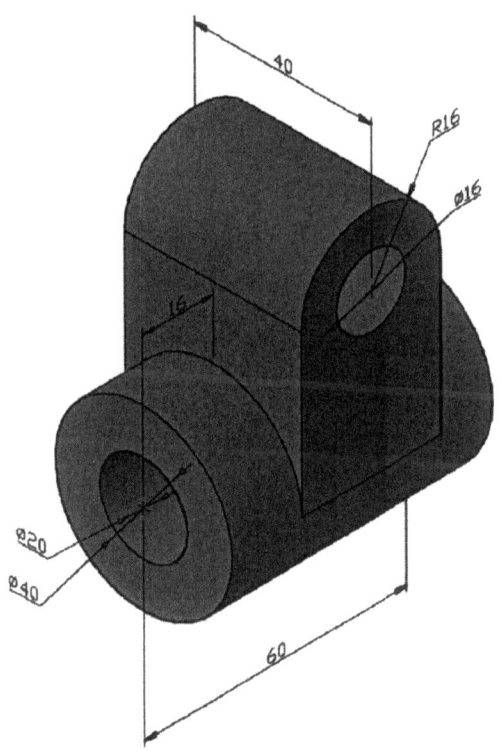

Draw the necessary views to represent the object using AutoCAD.
All the dimensions are in mm.

TITLE:	
DRAWN BY:	
CHECKED BY:	
SCALE:	DATE:
SECTION:	DRAWING No.:

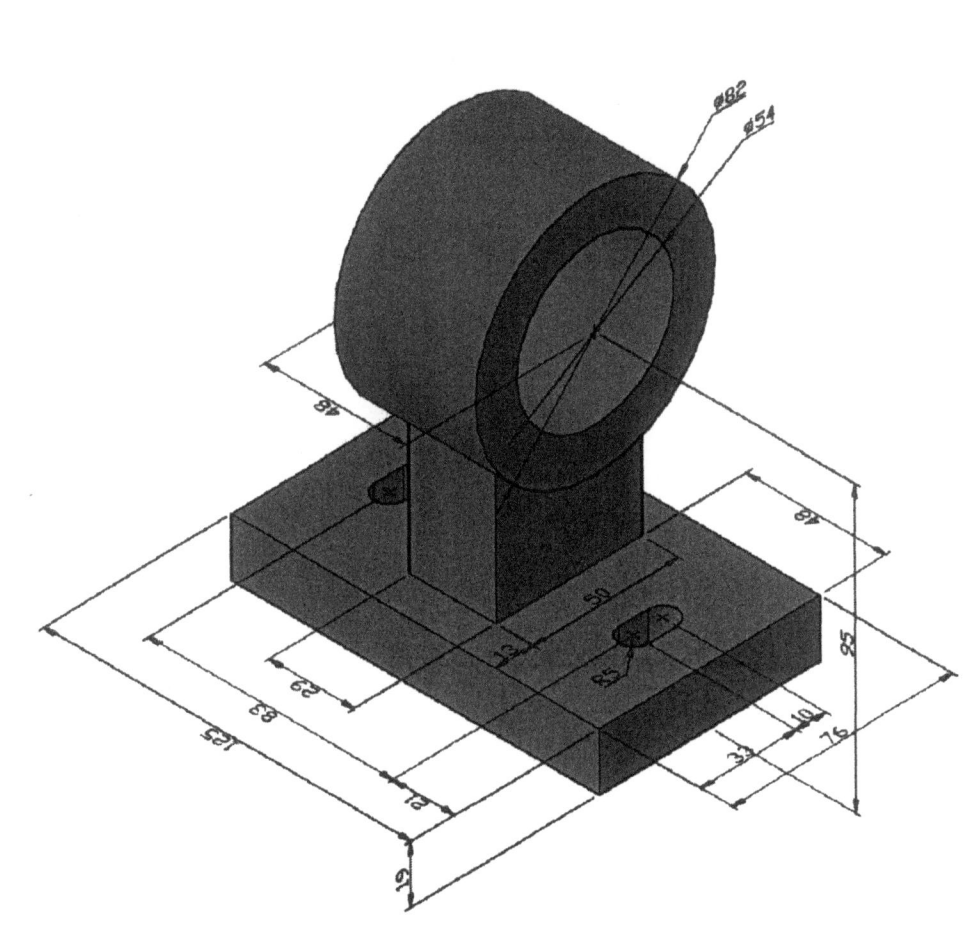

Draw the necessary views to represent the object using AutoCAD.
All the dimensions are in mm.

TITLE:	
DRAWN BY:	
CHECKED BY:	
SCALE:	DATE:
SECTION:	DRAWING No.:

TITLE:

DRAWN BY:

CHECKED BY:

SCALE: **DATE:**

SECTION: **DRAWING No.:**

TITLE:	
DRAWN BY:	
CHECKED BY:	
SCALE:	DATE:
SECTION:	DRAWING No.:

TITLE:	
DRAWN BY:	
CHECKED BY:	
SCALE:	DATE:
SECTION:	DRAWING No.:

TITLE:	
DRAWN BY:	
CHECKED BY:	
SCALE:	DATE:
SECTION:	DRAWING No.:

DRAWN BY:		DATE:
CHECKED BY:		DRAWING No.:
SCALE:	SECTION:	

TITLE:

DRAWN BY:	DATE:
CHECKED BY:	
SCALE:	DRAWING No.:
SECTION:	

TITLE:

| DRAWN BY: |
| CHECKED BY: |

| SCALE: | DATE: |
| SECTION: | DRAWING No.: |

TITLE:

TITLE:
DRAWN BY:
CHECKED BY:
SCALE:
SECTION:
DATE:
DRAWING No.:

www.ingramcontent.com/pod-product-compliance
Ingram Content Group UK Ltd.
Pitfield, Milton Keynes, MK11 3LW, UK
UKHW050645050526
12271UKWH00021B/337